Praise for *Waking the Frog*

"Great book—very readable. Scary, but helpful and hopeful. It says all that's needed to be said. And so thoughtfully."
—George Butterfield, businessman and philanthropist

"In *Waking the Frog* Tom Rand shows us clearly that climate leadership is not an issue of right or left but of right or wrong. Tom is a refreshing exception—a business leader who speaks with clarity and passion about the looming climate crisis."
—Tzeporah Berman, environmental activist and author

"A brilliant analysis of why we are stuck in our collective response to climate change—and, more importantly, convincing recommendations for solutions and a path forward."
—David Miller, politician and CEO, WWF Canada

"This is the perfect read for people who are sick of the polarized sound bites that currently dominate the climate debate. It's relatable and non-threatening but also blunt and to the point. Tom's clearly an authority working within the system but inspires people to mobilize and make change. A great read."
—Kali Taylor, executive director, Student Energy

"A polymath at his best, Tom Rand gets it. Climate change isn't just about science—it's culture, psychology, economics, the works. *Waking the Frog* is highly readable stuff. Even funny at times!"
—David Buckland, artist and founder, Cape Farewell

"In *Waking the Frog* Tom Rand manages to do something that no other book I know of on climate disruption has done. He takes a multi-disciplinary approach—blending economics, politics, psychology, statistics, and business analyses—to thoroughly dissect why inaction on this massive problem is so pervasive and what urgently needs to be done going forward."
—Ed Whittingham, executive director, Pembina

"In this lucid, timely, and highly readable exploration of the climate crisis, Tom Rand details the predicament the world finds itself in and the insidious mind traps that make it so hard for us to work our way out of it."
—William Seager, professor, department of philosophy, University of Toronto

"One of the best things about the very good *Waking the Frog* is Tom Rand's relentless, fact-based optimism. It's impossible to read this thing and not come away thinking 'Gee, maybe this climate change puzzle is solvable after all!' Highly recommended."
—Rick Smith, co-author, *Slow Death by Rubber Duck*, and executive director, Broadbent Institute

"Tom Rand explains why we're stuck, and why we need to get moving and take up the challenge of climate change as soon as possible, and most importantly how and why we should engage the power of private sector capitalism to transform promising ideas into real solutions."

—Dr. John Nielsen-Gammon, regents professor and Texas state climatologist

"[*Waking the Frog*] is an important piece of work from one of Canada's great entrepreneurs and change agents."

—Ilse Treurnicht, CEO, MaRS Discovery District

"Intriguing and very readable—it zips along nicely and shows some impressive scholarship. Well done!"

—Dominic Geraghty, executive-in-residence, EnerTech Capital Partners, ex-director of R&D,
Electric Power Research Institute (EPRI)

"*Waking the Frog* is an engaging, informative, and interesting wake-up call for a world that is sleepwalking towards disaster. It's also a surprisingly readable book, without undermining the seriousness of the topic."

—Andrew Heintzman, author and financier

"Take Tom Rand's deep knowledge and erudition on climate change and mix in Tom's inimitable style and you're left with a read that is energizing, highly informative, and entertaining all at once."

—Alex Wood, senior director of policy and markets, Sustainable Prosperity

"Tom Rand's clarion call to us, his fellow frogs, could not be clearer: 'Let's stop sitting here simmering and do something about it!' As an experienced entrepreneur and venture capitalist, he shows how practical, achievable measures can turn the climate threat into a promise."

—Walt Patterson, associate fellow, Chatham House, UK

WAKING *the* FROG

Solutions for Our Climate Change Paralysis

TOM RAND

ecw press

Published by ECW Press
2120 Queen Street East, Suite 200
Toronto, Ontario, Canada M4E 1E2
416-694-3348 / info@ecwpress.com

LIBRARY AND ARCHIVES CANADA CATALOGUING IN PUBLICATION

Rand, Tom, 1967–, author
Waking the frog : solutions for our climate change paralysis / Tom Rand.

ISBN 978-1-77041-181-4 (BOUND).
ALSO ISSUED AS: 978-1-77090-524-5 (PDF); 978-1-77090-525-2 (EPUB)

1. Climatic changes. 2. Climatic changes—Public opinion.
3. Climatic changes—Economic aspects.

I. Title.

QC903.R36 2014   363.738'74   C2013-907761-8   C2013-907762-6

Cover design: Michel Vrana
Images: frog on front cover © kazoka/shutterstock;
frogs on flap and spine © alptraum/iStock
Typesetting and production: Lynn Gammie
Printing: Friesens 1 2 3 4 5

The publication of *Waking the Frog* has been generously supported by the Canada Council for the Arts which last year invested $157 million to bring the arts to Canadians throughout the country, and by the Ontario Arts Council (OAC), an agency of the Government of Ontario, which last year funded 1,681 individual artists and 1,125 organizations in 216 communities across Ontario for a total of $52.8 million. We also acknowledge the financial support of the Government of Canada through the Canada Book Fund for our publishing activities, and the contribution of the Government of Ontario through the Ontario Book Publishing Tax Credit and the Ontario Media Development Corporation.

PRINTED AND BOUND IN CANADA

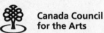

This book is dedicated to everyone who stays hopeful while working to move the needle on carbon. You inspire me.

CONTENTS

"There are no passengers on spaceship Earth, only crew."
— Marshall McLuhan, Canadian philosopher

The arrow is in flight. As I write these words in late 2013, the Philippines is trying to recover from Typhoon Haiyan and the U.S. remains gripped in the worst drought in living memory. Sure, we'll bounce back. Maybe next year will be great. But we sense what's happening. Changes are afoot. They're not good, and they've only just begun. This year's droughts, fires, and storms are but an appetizer for what the climate has in store. We can't stop it, but we might just be able to slow it down.

Progressives tend to see climate change as a real threat and conservatives do not. This book is for both. I hope the fact that I am a capitalist who operates within the system I'm critiquing makes those criticisms harder to dismiss. I'm not an outsider looking in but an insider looking ahead.

Since selling a global software company in 2005, I've dedicated my capital and time to moving the needle on carbon. When I built Planet

Traveler, the lowest-carbon hotel in North America in 2010, I did it partly to show the business community what can be done with our buildings — today and at a profit. I invest in emerging clean-energy technologies not just because I believe the market economy delivers rewards commensurate to the problems you solve, but because I passionately believe it's the most important work I can do. I decided to speak and write publicly on the carbon issue because, for reasons I didn't fully understand at the time, I wasn't seeing many of my peers in the business community speaking openly about what's at stake with our changing climate. The late, great Ray Anderson — American carpet magnate turned eco-evangelist — was a wonderful exception. There are others, like Bullfrog Power's Greg Kiessling, but most business leaders have yet to step up to the plate.

Captains of industry, pundits, and other civic leaders don't need to know the science (although it doesn't hurt). You need only make the reasonable judgment that NASA, the collection of National Academies of Science, and the International Energy Agency are more credible than Fox News or groups funded by billionaire industrialists, such as the Koch brothers. Our most august institutions are ringing the alarm as loudly as they can.

I write a lot about the United States in this book. Without the moral, financial, political, economic, and intellectual leadership of the U.S., the rest of us cannot possibly address climate change. With the U.S. absent, it's like rowing with half the oars out of the water and no coxswain to call the timing. America, we need you.

My own country, Canada, is equally absent from the scene. We're small but have a proud history of acting on the important issues of our times: the Second World War, peacekeeping missions, the ozone layer, the basis of the Right to Intervene — we've consistently punched above our weight. Our reputation as a nice (but tough) country was deserved. Those days are long gone. As a moral light Canada has faded.

We're a scourge on the planet. We silence our scientists and kneecap our finest environmental groups (including the National Roundtable on the Economy and Environment — hardly tree-huggers). We provide comfort to countries that prefer to dither by refusing to act ourselves. We were the first to withdraw from the Kyoto Protocol, the U.N.'s international treaty to reduce greenhouse gas emissions.

Current federal leadership argues that Canada's tiny two percent of global emissions makes no difference. We could use that same argument to talk ourselves out of voting. This is a churlish position, devoid of moral leadership or vision. It is also economically short-sighted. If Canada captured two percent of the global clean energy market, by 2020 our cleantech industry would be larger than our aerospace or automotive sectors. We live in a market economy. Those who solve big problems earn big rewards, and climate change has become the mother of all problems.

We haven't yet had a mature conversation about climate change in North America. The one about how the changing climate is likely to bring our stable global economy to its knees. Throughout this book I refer to *climate disruption* rather than *climate change*. The term helps circumvent the nonsense that this warming is part of a natural cycle and emphasizes our contribution to the coming changes and the speed at which they are approaching. I offer solutions along with criticism; the solutions here are less technical and more ways of knocking us out of our stupor. (I've written extensively about technical solutions elsewhere, see *Kick the Fossil Fuel Habit*.) The goal of this book is to make simple observations that might motivate you to do what you can with what you've got to stop our mad gallop to a very, very hot place.

My own hope lies in humanity's tendency to pull together when disaster strikes. The average American's response to Hurricanes Katrina and Sandy — like that of Britons to the floods that coursed through their towns, Albertans to their recent deluge, and Australians to their own floods, fires, and droughts—is to act with courage and humanity. When we are humbled by nature, we look to each other with humility. As climate disruption begins to bite and the culprit is clear, I believe we'll act with determination. I have to, because there is no other choice.

Tom Rand, P.Eng., Ph.D.
Toronto, Canada

Frogs, Hot Water, and Us

*"Man is condemned to be free; because once thrown into the world, he is
responsible for everything he does."*
— Jean-Paul Sartre, French philosopher

"If we don't change direction soon, we'll end up where we're heading."
— World Energy Outlook 2011, International Energy Agency

Every school kid has heard a version of the gruesome story of the frog,
the pot of water, and the stove. Put a frog in a pot of water, place the
pot on a stove, and turn on the element. As the water heats up, the frog
sits there until the water gradually boils the poor creature alive. The
story goes that cold-blooded frogs are not designed to respond to that
sort of temperature change. It's too slow to trigger panic or movement,
but it is fast enough to trap him. There must be some moment when
that froggy's nervous system lights up with the knowledge that something really bad is happening, but by then its blood has warmed to the
point that its muscles are useless. So it sits, paralyzed. The story is not
literally true—it's a metaphor, but a strong one.

Are we that frog? The question seems hyperbolic, alarmist. I am
certainly being deliberately provocative. But is it so crazy to ask? Each
time a presidential candidate proudly disavows climate science or
an energy company announces yet another massive fifty-year capital

investment in fossil-fuel development, our paralysis feels real. Each time extreme weather breaks new records, from raging fires in Australia to heat waves and intensified droughts in the U.S., our pot gets hotter. Even the pine beetle, scourge of Canada's forests and timber industry, is aided and abetted by a warming climate. Yet it's happening so slowly we look the other way. All signs point to continued paralysis in the very teeth of a storm of increasing intensity.

Waking the Frog is about how we can turn down the heat. We still have a small window of opportunity to act, but until we recognize our paralysis and ask hard questions about why we're paralyzed, we'll continue to stew. This book seeks root causes, not surface appearance. It may seem negative at times. That's because we have to take a good, hard look in the mirror. The first step to change is acknowledging who we really are.

Our paralysis is obvious. Given the state of play in the political and business sections of our daily newspapers, blogs, and television newscasts, it's clear we're not going to meaningfully grapple with carbon emissions any time soon. While we dither, argue, and delay, carbon counts rise well past the danger zone. We maintain investment levels in high-carbon infrastructure that last for decades. The economic, political, and sheer physical momentum of the carbon machine is enormous—and getting bigger. Efforts we make are largely symbolic. The degree of certainty in the scientific community about climate disruption has little effect on this dynamic.

And things are getting alarmingly hot. It looks less crazy each day to lose sleep over the heat-death of much of the planet and the collapse of our economy. Even the highly conservative International Energy Agency (IEA) recently confirmed we're on track for a 6°C (11°F) global average temperature rise this century.[1] That's a very hot pot! There's little room for error in our food system. The intensified droughts that 6°C (11°F) of warming bring will quickly dry up international grain markets. Our bountiful oceans are heating up (and acidifying from dissolved carbon dioxide) fast enough to disrupt the entire marine food web. A hotter atmosphere means lots of energy for extreme weather events. The list goes on. There's little chance we can adapt to such abrupt and dramatic changes—certainly not at anywhere near current population levels—however ingenious we may be.

The upshot is that while we have been waiting . . . and talking . . . and dithering, climate change has evolved from a somewhat abstract threat to a real and present danger. We're in hot water. Right now, we look a lot like that poor frog.

The good news is that we can solve the climate problem. The capital we need sits in our pension funds and money markets, the policy tools to unlock it are well understood (if politically problematic), and existing clean technology and emerging innovations are fully capable of powering our civilization.[2] Aggressive action is nowhere near as expensive as opponents claim. We can turn down the heat. We don't have to be that frog.

You'll find here some of the usual suspects: vested interests defending the status quo; money distorting policy and politics; the complexity of climate science; the ease with which critics can sow doubt; powerful free-market ideologues and rich industrialists, who fund anti-scientific nonsense; the fact that climate change happens over the long term but we think and act over the short term. It's a classic tragedy of the commons: the problem is global but action is local. Meanwhile, fossil fuels are cheap, abundant, and extremely useful, while clean energy is (for the short term anyway) more expensive and less convenient. All these circumstances contribute to inaction, but they are not enough to fully explain it.

I'm more interested in how it all hangs together—the rules of the game, not just the bad actors. The global economy is shaped, broadly speaking, by democratic, free-market capitalism.[3] My concern is that there seems to be something wrong with the structure of the economic machine we've built: democracy, free markets, and capital interact in ways that *exacerbate* paralysis on a global challenge like climate change. Mere malfeasance, laziness, or vested interests are not enough. Something deeper is at play. I hope to expose structural constraints — the equivalent of whatever it is in the frog's DNA that lets it sit idle as the water heats up.

Democracy dictates that the public endorse the direction we take. It's only through our collective permission that priorities are shaped, policy takes form, and laws come into being. Yes, capitalism's monetary lifeblood allows for all kinds of political distortions, but it's the court of public opinion that politicians ultimately buy and sell. We the people, for better or worse, must decide what action to take on climate change. But we are not demanding action. Why not?

Psychology and cognitive science have a lot to say about that. Denying or ignoring a problem is perfectly natural — it's our default reaction. It takes cognitive effort to see a problem for what it is, despite the overwhelming evidence and expert opinion before us. Denial of climate disruption, which I'll refer to as climate denial, can be subtler than outright scepticism about the reality of climate change. A softer form admits a hotter pot, but it tempts us to ignore how serious a fix we are in. Denial is a siren song, indeed, but it's one we can learn not to heed.

By capitalism I mean the uniquely bare-knuckled U.S. style of free-market capitalism that drives the emerging global economy, in which international capital flows tend to dominate domestic political agendas — including Canada's. Today's global economy is the most powerful and creative social tool in history. The capital that sloshes around global markets, seeking to maximize risk-adjusted returns, marks a greater measure of wealth than humans have ever commanded. Redirecting investment away from the highly profitable fossil fuel sector and into low-carbon infrastructure requires massive intervention in the market.

I see no evidence that free markets can address climate change. The proven reserves of the global energy giants—which sit on their balance sheets as assets to be sold—are already four times more than what we can safely burn. The market is telling us, loudly and clearly, that unless the rules are changed, climate change will win. Market freedoms trump intelligent action on carbon time and time again. Yet the free market is itself a myth—an academic fiction built on the mathematics of a bygone era. A modern, dynamic view of the economy, one that acknowledges its complexity and unmatched creative potential yet provides effective tools to tame it to our needs, is a more useful model for the twenty-first century.

Our metaphorical hot water is not a political issue but a practical problem. Both right and left are in the same pot, after all! A compromise recognizes the role a revitalized market economy might play: neoconservatives admit it must be tamed, and the left acknowledges its unmatched creative power. If both sides leave their dogma at the door, we can come together on the basics of a solution. Only a price on carbon can simultaneously harness and unleash the most powerful tool we have at our disposal: the market.

Striking deeper at the heart of traditional conservatism is a paradox: business as usual will not preserve the status quo. At the core of conservative thinking—and I refer now to the small-c conservative in all of us—is the idea that we tinker radically with our social, economic, and political structures at our peril. Incremental change is preferred, and conservative thinking has always been a steadying hand at the tiller. But we now face the hard reality that incremental changes to our energy economy will take us over a cliff and onto the rocks below. We all must somehow endorse radical change to preserve our way of life. That paradox is difficult to resolve, and many people choose to wilfully deny our best science as a way out.

Economics plays a central role in this story. In the guise of cold, sober cost-benefit analysis (who disagrees we need that?), woefully inadequate economic models are used to pass judgment on the net costs of climate action. These models use a fatally flawed interpretation of climate science, dumbing it down so that they are nearly useless. They arbitrarily impose estimates of the economic damage our warming pot will bring (a "damage function"), with no decent empirical or theoretical basis whatsoever. Despite a history of predictive failure, they churn out precise numbers meant to capture climate consequences that are decades distant. This unjustified precision betrays a dangerously false confidence, but they confirm what we *want* to hear: relax, it's not so bad.

A traditional cost-benefit analysis is great for modeling the net benefits of single projects in isolation, such as building a bridge or refurbishing a building, but it's unsuitable for the global, systemic changes coming our way. A hotter pot changes *everything*, including all those variables a model likes to hold constant (like growth rates). We use these models out of habit and for the falsely comforting story they tell. There are better alternatives. Economics can revitalize itself. New thinkers are more collaborative, more appreciative of the complex nature of the economy and the need for a multi-disciplinary approach that includes moral questions that cannot be reduced to traditional accounting.

But life feels grand at the dawn of the twenty-first century! Our collective affluence sits on a mountain of cheap and plentiful fossil fuels. Who in their right mind really wants the fossil fuel party to end? But end it must, and the scale of the job ahead is massive enough to invite paralysis. Staring at that huge mountain you're expected to climb before sundown can make you want to sit down, remove your hiking shoes, light a campfire, and open a bottle of wine.

An honest assessment of carbon reduction is fatiguing. It's not irrational to despair. But despair is a choice and not one we need to make. Previous generations bore down on some pretty serious problems. Our descendants should expect no less of us. Corny as it sounds, you climb a mountain by taking one step at a time. The trick is to start.

Yet many of us don't feel as if the risks of droughts, fires, and floods are ours to bear, not directly. So why start? Our collective affluence allows us to ignore the storms gathering on the horizon. It surely won't hit *us*, after all—that stuff happens to poor people on TV! Perhaps Hurricane Sandy has changed that. Time will tell. Climate change is a risk we all share, not just because Wall Street can flood, and the Prairies can dry out, but because grain markets are global and no country with modern weapons is willing to starve without a fight. There is no gated community that climate change can't infiltrate.

Our affluence hides another risk, one we must take if we're to quickly rebuild the largest piece of infrastructure humans have ever created. Energy flows permeate every corner of our economy. Fossil fuels power our civilization. Our professional classes have, for good reason, self-selected to be highly conservative when it comes to changing core infrastructure. Our legal and corporate structures reinforce that conservatism. Those with their hands on the levers of power—CEO, financiers, captains of industry, politicians—are paid very well to continue to do what they do. They are, explicitly and implicitly, rewarded to keep the ship on course. If you made a lot of money yesterday and it looks like you'll make a lot more today, why on earth would you stick your neck out, risk your professional reputation, and change direction? Why should a newcomer who advocates radical ideas like a massive restructuring of global energy flows be accepted to handle those levers of power? But the ship must be turned. We need a different sort of leadership now.

These forces do not act in isolation. Feedback makes the whole larger than that sum of its parts. Some are easy to identify. Vested interests take advantage of denial's siren song. Peddling seeds of doubt with pseudo-experts works so well because those seeds land on fertile psychological soil. A public that thinks the science is in doubt doesn't demand, or even endorse, meaningful policy or radical change. The result is political inaction and an entrenched status quo.

The status quo's appeal is amplified when the market allows those who have accumulated sufficient capital to use their huge swaths of media as a kind of megaphone. When Fox News gets away with calling climate change a "scam" on nightly newscasts and weather announcers—trained mainly to look at the camera and smile—are trotted out as "experts" to counter the weight of unequivocal statements of alarm by more than a dozen international equivalents of the National Academies of Science[4], the public interest is no longer served by the private ownership of media. Yet that is precisely what the market allows, to say nothing of the role of money in Washington or the current anti-science oil-patch salesmanship in Ottawa.

If this seems too severe, recall that the U.S. has presidential candidates who must repudiate climate science if they are to have a chance at getting past the primaries. And while Texas dries out in the most severe drought since record-keeping began—which is precisely what climate scientists have predicted—Texas governor Rick Perry prays publicly for both rain and an end to the Environmental Protection Agency. That's like praying for your car to get fixed and for the mechanic to get sick all at the same time. Closer to home, our own federal minister of the environment, former broadcast journalist Peter Kent, often seems to be nothing more than a pitchman for oil pipelines. Harper's recent omnibus budget bills, which gutted anything with even a faint whiff of environmental stewardship, were more than mere clearing of red tape. The result is the climate gridlock we see.

But we are very *unlike* that poor frog, of course. We are conscious and capable of choosing our future. We are clever and plan and act over large periods of time and space. We put a man on the moon, for goodness' sake! We split the atom and have observed events in space so distant they represent the dawn of the universe. We routinely drill kilometers into the earth for oil in the extreme conditions of the open ocean. We certainly *can* turn down the heat; it's a matter of collective will and determination.

Our paralysis is one of choice. The required capital, technology, industrial base, and policy tools all exist. New innovations in energy storage and production emerge constantly, as do innovative financial tools and means of public engagement. So the relevant question is not "can we turn down the heat?" but "will we turn it down fast enough to avoid economic collapse?" I believe this is the most fundamental economic and political challenge of the twenty-first century. I hope we can, of course, but hope is a poor cousin to deep understanding of the causes of our paralysis. We must first take the world as it is before we can hope to change it.

The great existential philosophers of the twentieth century appreciated the degree of freedom we have to make change and the depth to which that change may go. It is consciousness that most uniquely defines us and is our source of meaning and possible future. Only the human animal, with our capacity for self-reflection and explicit sense of free will, can engage in the act of remaking him or herself. In so doing, we forge new meaning and choose anew how our future might unfold. Only when we take responsibility for our freedom will we be able to chart a course through the dangerous waters ahead. It all starts with the individual choice that lies at the heart of our democratic capitalist system.

Democratic free-market capitalism is the most powerful political and economic system in history. It has such a pull over time and geography that American political scientist Francis Fukuyama saw it as the necessary endpoint of human history. Fukuyama believed all countries would embrace it, and he saw us as quite literally at the *End of History*. But he was wrong. It is yet to be determined whether those powerful forces will be tamed and brought to bear against what is fast becoming our endgame: the war against carbon. The real test of the free market was not Soviet Russia nor is it Communist China. It's climate disruption.

From Serious Science to the Theater of the Absurd

"Climate change . . . could alter the way we live in the most fundamental way. . . . It is life itself that we battle to preserve."
— U.K. Conservative prime minister Margaret Thatcher to the U.N. General Assembly, 1989

"President Obama promised to begin to slow the rise of the oceans . . . and heal the planet."
— Mitt Romney, U.S. Republican presidential candidate, to derisive laughter at the Republican convention, 2011

OFF TO THE RACES

The battle to save the planet started out well enough. When U.K. prime minister Margaret Thatcher addressed the United Nations General Assembly in November 1989, there could not have been a more credible advocate for strong action on climate disruption. The world seemed poised to act, and leading the charge was none other than the Iron Lady herself—champion of the free market, military hawk, and good friend of U.S. president Ronald Reagan. Fast-forward twenty years. Today, an outpouring of climate scepticism is required to establish bona fide conservative credentials. From former U.S. governor Mitt Romney (Massachusetts) to former governor and vice presidential candidate Sarah Palin (Alaska) and from Fox News to the American

Enterprise Institute, neoconservatives have somehow positioned a scientifically illiterate denial of climate disruption into the national debate. Even when the problem is acknowledged, it's played down as more of a nuisance than a threat. What on earth happened?

Whatever anyone thought of Thatcher and her policies or priorities, the last accusation one could aim at her was being a shill for the environmental movement. Thatcher was as vigorous a defender of the free market as anyone on either side of the Atlantic. As a Conservative prime minister of Britain, she oversaw one of the most dramatic overhauls of the British economy in history. Tearing back regulations, opening up markets, massively lowering tax rates on business and the wealthy—Thatcher brought Britain kicking and screaming into the free-market economy that it has today. Thatcher was a kind of trans-Atlantic sister in arms to then U.S. president Ronald Reagan. Both were driven by similar ideologies: the free market was how individuals could best express their freedom and how corporations could best build economic strength. The market would conquer all.

My family was living in central England during the tumultuous years of 1981 and 1982, while my father, a professor of molecular biophysics, spent his sabbatical studying at the University of Nottingham. I had a paper route, an early outlet for my own entrepreneurial energy. On my twice-daily journey among the townhouses of Nottingham, I would read on the front pages of the national papers about the brutal and highly confrontational coal miners' strike just to the north of us. Thatcher had made her mind up to break those unions, and she did. The reasons were not environmental. Britain was to embrace nuclear power, and Thatcher was determined that no single industry would stand in the way of the larger national interest. That same steely determination would emerge time and again during Thatcher's eleven-year reign—during the riots over the Poll Tax, the Falkland Islands conflict with Argentina, and when confronting her political opponents in Parliament. Whatever you may think of Thatcher's politics, she was damn good at getting things done once she set her mind to it.

So when the Iron Lady put her reputation on the line to warn the U.N. General Assembly about the dangers of climate disruption, it seemed environmentalists finally had the ally they needed. Here was a political leader with strong neoconservative credentials speaking about an environmental threat to civilization, one that was caused by humans and was urgent enough to warrant global action at the level of the U.N. The stakes were high, as she told the assembly, "It is life itself that we battle to preserve."

A statement like that from the conservative side of the aisle is totally out of place in today's political environment. You don't earn your neocon stripes these days without denying the dangers, causes, and very existence of climate disruption. But Thatcher operated in an atmosphere in which scientific associations such as the Royal Society and the National Academies of Science, and scientists themselves, were held in higher regard. Perhaps more importantly, Thatcher herself was scientifically literate, and, even in the 1980s, it was clear to anyone with a degree of scientific knowledge that danger lurked on the horizon.

It was known that carbon emissions, unchecked, could and would wreak havoc on our civilization. We didn't know exactly when, but we were certainly headed for stormy weather, so Thatcher rang the alarm.

There was a long history to the science by this time, and Thatcher's position was not as controversial as it might seem today. We'd known for more than a hundred years that carbon dioxide is a greenhouse gas that traps the sun's heat like an insulating blanket. It is carbon dioxide that is largely responsible for the "Goldilocks" temperature of our planet. Without it and water vapour, the average temperature would be a bracing -17°C (1°F)! But there can be too much of a good thing.

In 1824, French physicist Jean-Baptiste Joseph Fourier described how the atmosphere is asymmetrical when it comes to thermal energy. That is, the gases the atmosphere contains are less transparent to outgoing thermal energy than they are to incoming shortwave energy from the sun. That marked the birth of what we call the greenhouse effect. In 1859, Irish scientist John Tyndall experimentally confirmed the different asymmetric properties of various gases. Calculations as to how sensitive the climate might be to various levels of carbon dioxide began with the Swede Svante Arrhenius in the late nineteenth century.

By the 1950s, the speculative became the empirical. American scientist Charles Keeling began measuring carbon dioxide levels with great accuracy from a mountaintop in Hawaii. Keeling's readings were so accurate that we could see for the first time the world's forests "breathing" on an annual basis: leaves falling from the dominant northern forests increased carbon levels, and spring growth reduced them. But the overall level was clearly rising. Global temperatures were also being recorded. Seen over time, by the 1980s the correlation between rising temperature and rising carbon was unnerving. Further, the rate of temperature rise was in the ballpark of what scientists predicted from the rising carbon.

With the advent of computing technology, climate science evolved further. Once merely theoretical then confirmed with empirical measurements, climate science became predictive, with scientists basing their findings on ever more complex atmospheric models. These models didn't just measure the amount of incoming and outgoing energy; they took into account many of the interactions that regulated the flow of carbon from sea to sky, from sky to tree, and so on. By the late 1980s, these predictive models were delivering some very bad news. The climate might not just warm incrementally with added carbon dioxide, but it could shift very abruptly to a new, and very probably hostile, equilibrium. Those early models, clunky as they were, were on the right track. Worse, many predictions—such as the melting of the Arctic sea ice—are happening much faster than originally thought.

It all seemed to come to a head in the late eighties. My hometown of Toronto hosted the first major intergovernmental conference on climate disruption in 1988—coincidentally, the hottest year ever recorded in the 130-year history of global temperature record-keeping. The year before, world leaders had met in Montreal to deal decisively with the ozone problem. A repeat performance seemed in the cards: scientists' warnings would prompt leaders to hash out binding agreements in a coordinated response to a global problem. The Toronto event ended with a call for a twenty percent reduction in the carbon emissions of industrialized economies from 1988 levels by 2005.

Activity on the climate front reached a fever pitch in 1988. The World Meteorological Organization approved the formation of the Intergovernmental Panel on Climate Change (IPCC), later backed by the might of the United Nations, with a mandate to provide consensus-based, top-tier advice on the science. NASA scientist Jim Hansen addressed Congress in the midst of a serious drought in the U.S. Midwest, famously saying: "It's time to stop waffling . . . the greenhouse effect is here." When *Time* magazine named an endangered Earth "Planet of the Year," the general public became fully engaged in the discussion.

So by the time Thatcher gave her speech at the U.N., a strong lineage to climate science and a lot of government interest existed. From Fourier in 1824 to Hansen in 1988, we had come a long way in our understanding of what was happening and what was at stake. Governments of industrialized nations had committed to reducing emissions and had formed a globally recognized body of scientists from whom to take formal advice. Thatcher's address, while seemingly alarmist by today's standards, was merely another domino to fall in what seemed to be an inexorable march to action.

In response, other senior public figures inevitably lined up to express concern. Then U.S. president George H.W. Bush went to the Earth Summit Rio in 1992 to affirm that the United States was taking the threat seriously. That same year, Al Gore published *Earth in the Balance* and was then elected vice president under Bill Clinton. The U.S. was poised to join Britain in taking a leading position on the issue, bringing its industrial, financial, and entrepreneurial might to bear. With the Kyoto Protocol, the entire world came to the table with the first global agreement to reduce carbon emissions. The developed world would lead, and the developing world would follow.

Things were looking good. Words mean a lot less than action, of course, and political promises are much easier to make than to keep. The ozone deal, while showing the possibility of generating global cooperation toward a common threat, was much simpler and less costly to implement than a similar deal on carbon. Carbon emissions permeate every sector of our economy, while ozone emissions affected a very thin slice of the economic pie. Real work on the climate could only begin with the difficult job of executing deep changes in the way the global economy found, used, and paid for energy. Nevertheless, we could see a path forward. As the saying goes, where there's a will, there's a way — and we certainly seemed to have the will.

To solve what was arguably one of humanity's greatest problems, thoughtful policy would be informed by the best of the world's peer-reviewed scientists. That policy would unleash the full power of free-market capitalism: the innovation of entrepreneurs, the intellectual depth of the world's finest research organizations, the huge scale of available capital, diligent bankers, the risk-taking culture of the venture capitalists, and the might of the big energy firms. Democratic free-market capitalism seemed not just capable but uniquely capable of solving the carbon problem.

The Theater of the Absurd

But it's all gone sideways. More than twenty years later, we pump out carbon at ever-increasing rates. Global capital investment in high-carbon energy infrastructure has barely paused, even with the beginnings of the global recession in 2008. The political will to create strong market signals to curb emissions has evaporated. Kyoto is a shipwreck: the United States never ratified it and Canada was the first to formally drop out in 2011. Even as the most respected scientific organizations in the world deliver ever more strident warnings of global economic and ecological collapse, Republicans in the U.S. still earn their neocon stripes by calling climate disruption a scam, and our own prime minister can barely utter the words. Public dialogue has devolved into a kind of theater of the absurd.

It's true there are pockets of reasoned discussion in academic, political, and even industrial sectors. Reasonable disagreements about how to best solve the problem are part of the inevitably messy dynamic of a healthy democracy. Deciding upon and executing policy; balancing priorities over the short and long term; delivering equity between different countries, generations, and socio-economic classes; dealing with the fallout of fading industries like coal — these were always going to involve robust discussion, but talk about solutions is a distant whisper in today's echo chamber of television, blogs, and newspapers. Even reasonable discussion does not acknowledge how bad this storm is going to be. That's not what the broader, dominant public and political arguments are about. And boy, do we argue.

Some claim, in the very teeth of a warming and ever more hostile atmosphere, that climate disruption is simply not real, that the Earth is not getting any warmer. The rise in temperatures, reported in multiple data sets obtained by independent laboratories, is decried as a kind of statistical illusion, sleight of hand, or even outright fraud, as if there weren't robust and objective statistical analysis methods that can separate the signal from the noise and long-term change from short-term variance. Small human failings among the thousands of climate scientists are seized upon as evidence of this fraud, while the mountain of evidence untainted by scandal is ignored.

Others acknowledge temperatures are rising but don't accept that our carbon emissions are the cause. These folks are often reduced to blathering on about sunspots and cosmic rays. When it's pointed out to them that we have satellites to measure solar activity and that current warming is not correlated to an increase in activity, the result is often not more than a shrug, as if such measurements weren't relevant or the satellite data are part of a larger global climate conspiracy.

The latest fashion in denialist circles is to point out previous, naturally occurring changes in the Earth's climate, as if the existence of natural variations precludes the possibility of human-induced warming. It does no good to point out that these earlier periods of warming and cooling were caused by Malinkovitch cycles, which are well understood and very predictable (if subtle) changes in the Earth's orbit. The grounds of the argument then shift to a kind of fatalistic acceptance: "We've been through these sorts of things before," goes the common refrain, "so we'll go through it again." That's akin to saying forest fires occur naturally, so we needn't worry about all those nasty people pouring gasoline everywhere.

To a scientifically untrained public, these positions might sound reasonable. Not everyone has studied statistics, and it's easy enough to change the timescale of a temperature chart to make it look less dramatic. Changes in solar irradiation do alter temperature, as do the long, slow variations in the Earth's orbit. Why not assume this is more of the same? Most people are too busy living their lives — raising kids, going to work (if they're lucky enough to have a job), making mortgage payments—to bother about such specialized knowledge. The public is easily fooled, and it's not their fault.

Our business leaders, politicians, and pundits should know better. It takes very little work to see past these silly arguments. Those who affect public debate should know of what they speak, particularly when what's at stake is nothing less than our continued wealth, prosperity, and even survival. You don't even have to study the science, only to have the minimum wisdom required to recognize who is credible on the issue and who is not. When the cacophonous blogosphere makes one claim and a peer-reviewed journal another, when an out-of-touch British peer with no background in science contradicts the findings of NASA, or when conservative media pundit Ezra Levant is at odds with the National Academies of Science, it's not a stretch to expect our civic leaders to know who should be given the benefit of the doubt. If we demand truth in advertising, can we not demand the same in scientific commentary?

It is somehow expected that we should even deny our own senses. We see evidence all around us, in subtle changes from year to year in our own backyards and in an alarming number of news stories about extreme weather. Our deepest intuition tells us the rash of droughts, floods, fires, and hurricanes point to something ominous. That intuition happens to be right.

Clearly no single weather event can be explicitly linked to climate disruption, just like no one pixel makes a picture and no one note a melody. Weather is not climate. The drought in Texas, the ongoing wildfires, the droughts and floods in Australia, the extraordinary heat wave that temporarily shut down Russian grain exports in 2008, the devastating flooding in Pakistan, the melting of Canadian Arctic ice and permafrost—none of these can be said to be exclusively "caused" by global warming. That's not how climate disruption works.

But just as you can distinguish an image once you're given a few more pixels, we are seeing the ugly face of climate disruption emerge. It's not just that we can see an identifiable pattern. Crucially, this pattern of increased extreme weather matches the predictions of our best climate models. The sort of violent atmospheric activity we see before us is precisely what is expected when the atmosphere contains more thermal energy. It's becoming ever more irrational to ignore the connection.

Climate disruption causes extreme weather in the same way obesity causes heart attacks, cigarettes cause cancer, and steroids create better home-run hitters. The relation is statistical. Climate disruption increases the probability and severity of extreme weather events. The Texas drought is exacerbated and made more likely by climate disruption. Enough of these statistically abnormal weather events are an indication that climate disruption is here. It is no longer an abstract theory about something that may happen one day in the future. It's an empirical fact. It's like an uninvited guest has crashed our economic party, and instead of working together to eject the beer-stealing, furniture-busting troublemaker, we're in the kitchen debating his very presence.

U.S. senator Jim Inhofe, for instance, makes the case for a vast conspiracy, stating bluntly in a Senate speech that climate disruption "is the single largest hoax ever perpetrated on the American people," insisting, against expert opinion and empirical evidence, that "there is no relationship between man-made gases and global warming."[5] Inhofe's words are echoed by doddering fools like British peer Lord Monckton, whose pedigree and background in scientific inquiry isn't much better than that of my young nephew, who is still on picture books and toy trucks.

Never mind that the degree of collusion required to pull off a scam of this magnitude is, to put it mildly, astronomic. Scientists separated by two hundred years would have needed to coordinate their observations and theories, and atmospheric chemists would need to coordinate their message with geological physicists. Data trails of physical evidence would need to be planted in ancient tree rings, lakebed sediment, and Arctic ice cores. Thousands of ambitious grad students, otherwise best advised to advance their careers by knocking dominant thinking on its head, would have to trade in those ambitions for mere collusion. I think the conspiracy is a bit difficult to accomplish, never mind that there is very little motivation to do so. The big money's always been in oil, not solar panels or climate science.

While American neocons set the batty bar for politicians, Fox News has clearly set the silly standard for the media. Repeatedly calling climate disruption a "scam," Fox has forced some of North America's most trusted scientific institutions, including the likes of the National Academy of Sciences, to take a back seat to talking heads whose only real training is to look nice on TV and read weather reports. A survey done by George Mason University found that more than a quarter of television weathercasters agree with the statement "Global warming is a scam," and a majority believe that, if warming is occurring, it is caused "mostly by natural changes."

That a weathercaster is supposed to know something about atmospheric physics on a par with NASA or the IPCC is like saying a high-school cheerleader can play basketball in the NBA. It's not just a difference in degree of skill or experience; it's simply the wrong area of expertise (a category mistake, as philosophers would say). Weathercasters don't study climate; they study how to say, with aplomb, "Tomorrow will bring clouds and a thirty percent chance of showers."

Fox has also produced some of the most vitriolic diatribes against any sort of market interference to solve the climate crisis. It holds market freedom on a high altar, a natural force to which all else must bend. It's the Chicago School, only dumbed down and on steroids.

The first time I heard American conservative media host Glenn Beck call business magnate and philanthropist George Soros a watermelon — green on the outside, red on the inside — I thought the often hysterical talking head had actually hit on a dangerously pithy expression. Dealing with carbon does require interference in the free market, that's clear, and so the environmental aspirations of one of the greatest capitalists of our time are quickly reduced to socialism in disguise. Any attempt to use the power of the market to solve the climate problem is instantly reduced to the moral equivalent of Lenin. Beck (who became too controversial even for Fox News) hit on a meme so disarmingly simple that it denigrates an entire breed of capitalist while hiding any number of inconsistencies. It's one of the best (and worst) sound bites I've heard.

People have always held strange beliefs, and one of the great things about democracy is that we are free to believe what we want, to talk about those beliefs, and, often, to act on them. Want to travel to an ancient Incan pyramid to recharge your energy crystals? Go for it. This is all part of the wonderful tapestry of human life in a democratic society. Strange beliefs are not harmful or something to sneer at, in and of themselves.

In contrast to energy-crystal charging in Peru, climate disruption denial brings direct harm to ourselves and others. Climate disruption incredulity comes at a cost. It does damage in much the same way my hitting you would — it's just more abstract and dissipated over time and space. Libertarians are loath to acknowledge climate disruption because they are loath to admit any restrictions on liberty. Freedom of belief ends when it does harm to others. Pointing out that my freedom to move my fist ends at your nose seems lost on libertarian ideologues.

U.S. President Barack Obama's top science adviser, Dr. John Holdren, has likened our situation to being in a car with bad brakes that is driving at top speed in the fog and heading toward a cliff. We don't know exactly where that cliff is, but we know it's there somewhere. Rational behavior would be to hit the brakes, but we're not reaching for them. We're not even arguing about how hard to hit them or when to hit them. It's like we're in that car, heading toward a cliff, and we're arguing about which radio station we should be listening to.

Twenty years ago, Margaret Thatcher warned the world of the dangers of climate disruption. From that moment of clarity, we have devolved into a paralyzing cacophony of bullshit. How did it happen?

Opinion, Anyone? Scepticism as Intellectual Vice

Many climate sceptics, from Canada's political commentator Rex Murphy or *Globe and Mail* columnist Margaret Wente to America's favorite ex-TV host Glenn Beck, sincerely believe their scepticism is a form of intellectual virtue. It is not. Their brand of scepticism is a vice.

Admittedly, climate disruption is much more complicated than just a statement about atmospheric chemistry and its effect on global average temperatures. There is more to it than just science. Atmospheric chemistry is a *thing*, and there is a complex causal chain between that thing and us—political, economic, and social repercussions; regional disparities; industries under threat; and shifting priorities, to name just a few. Climate disruption is more than a thing; it is an interdisciplinary phenomenon. A very complex dance begins as its effects and our responses start to interact.

Scepticism about climate disruption as a fact is very different from scepticism about climate disruption as an interdisciplinary phenomenon. Pundits like Murphy or Beck have confused the two as well as confused the very different roles that scepticism might play. Just as "theory" plays one role in everyday discourse (I have a theory about why Amsterdam is so friendly) and another within science (the force of gravity is part of a theory about the nature of physical mass), so too does scepticism play various roles in public discourse. Gravity is not just a theory in the same way my view about Amsterdam is just a theory. While scepticism about climate disruption as a *phenomenon* might be the legitimate purview of many disciplines, scepticism about climate disruption as a *fact* is not.

What I'm calling the theater of the absurd is primarily under-written by this confusion. Judgments on and opinions about climate science have evolved from being the jurisdiction of those whose professional lives are dedicated to the study of the subject to being the business of just about anyone with a soapbox, television show, or blog. The fact of climate disruption — an empirical question about atmospheric physics — has become a topic about which the fashionable commentator's opinion is supposed to matter.

Pundits are not generally sceptical about statements physicists working at the Hadron Collider, the world's largest and highest-energy particle accelerator, make about the nature of subatomic particles. It would be ridiculous to do so. They might be sceptical about the money spent on such a project or why the Hadron Collider is located in Europe rather than the U.S., and so on. That scepticism is directed toward the Hadron Collider as a phenomenon, and this is part of a healthy public debate. But scepticism about the nature of matter seems absurd.

Scepticism becomes a vice when it hammers away at a broad consensus of expert opinion that is warning of an existential danger. The policy commitments that climate science demand need broad public support. Sceptics sap that support without intellectual justification.

For Rex Murphy, public acceptance of expert opinion on climate disruption amounts to indoctrination. Margaret Wente asserts that climate cannot be controlled by human behavior and that climate change has a "PR" problem because of a supposed pause in warming. For Glenn Beck, climate disruption is a communist conspiracy. U.S. senator Jim Inhofe thinks it's a scam, pure and simple. These sceptics explicitly cast themselves against the orthodoxy of our time, as noble knights who are standing up to society's pressure to conform.

This is nonsense. Climate disruption does not have a "PR problem." It has a problem with scientifically illiterate pundits passing judgment on a topic for which they have no expertise. Discussing climate disruption as fact is not like discussing politics or art. The opinions of laypersons are not relevant. It's hard science, and the truth of the matter has been settled by those qualified to make the judgment. Though often amusing, the denialist racket delays our participation in a new economy and makes it more difficult to have the public and adult conversation we so desperately need—the one about how volatile nature has become and how bad it will get.

But there is more to this theater of the absurd than dumb reporting—a lot more. These seeds of doubt are dangerous because they land on fertile psychological soil and are amplified in a giant media-driven echo chamber built by well-funded business interests that are working very hard to maintain the status quo. That's not healthy debate; that's manipulation of the public interest.

Dollars and Sense — Theater Gets Serious

Occasionally, the conversation moves past socialist plots and scientific conspiracies, but it remains absurd all the same. I've no bone to pick with economists. Theirs is a serious business, and there's no question that we'll need a detailed and comprehensive analysis before we can commit to policies to tackle climate disruption. But the assumptions behind their economic models—and, more importantly, the way in which their conclusions are explained to the public and the policy advisers who rely on their advice—are troubling.

Furrow-browed economists and captains of industry have consulted their spreadsheets and emerged with dark predictions of economic ruin and sputtering factories. If we were to commit to the emissions reductions those naive environmentalists demand, they say—even of the very mildest sort, like the Kyoto Protocol—economic ruin would ensue! We'd all end up living in cardboard boxes!

George Bush the Elder initially showed support at the Rio Summit (the precursor to Kyoto), but he made it clear within months of taking the White House that the seriousness of economics would trump the idealism of the environment. Writing to Senator Chuck Hagel, he said, "As you know, I oppose the Kyoto Protocol because . . . it would cause serious harm to the U.S. economy."[6] Later that spring, he released a press statement that stated, "For America, complying with those [Kyoto] mandates would have a negative economic impact, with layoffs of workers and price increases for consumers."[7]

Canadian prime minister Stephen Harper, a trained economist, once famously called Kyoto a "socialist scheme designed to suck money out of wealth-producing nations." He also said, "I'm talking about the 'battle of Kyoto' . . . the job-killing, economy-destroying Kyoto accord."[8] His views haven't evolved much since then, but his choice of language has—more as a reaction to looking silly than from any real understanding of climate science or the economics of our response, in my view.

Under Harper's leadership, Canada formally opted out of Kyoto in late 2011. The reasons given were economic: the cost in job losses and dollars was supposedly too much to bear. Even more absurd was the notion that we'd have to ship billions of dollars overseas. All that was patent nonsense. Kyoto had no teeth, and there were always mechanisms for us to avoid any real penalties. Kyoto was nothing more than a measure of the world's ability to cooperate, to see if we could act collectively—a test of who could play nicely together on the playground. Canada's withdrawal was like taking our ball in a huff and going home, refusing to play at all.

Meanwhile, in Australia, Prime Minister Julia Gillard recently pushed through a carbon tax as part of a plan to build the world's second-largest emissions-trading market. At great political risk and after a very bitter fight, the measure squeaked through Parliament. While its passage may portend a changing tide, furious opposition coalesced around Opposition leader Tony Abbott. Abbott, who has called climate disruption "complete crap," made it clear the fight was not over: "We will repeal the tax. We must repeal the tax. This is a pledge in blood."[9] He did repeal that tax, after being elected in 2013.

At first blush, this might not sound so unreasonable. Economic ruin might indeed be worth the hyperbole of blood pledges. As we seek deeper emissions cuts beyond Kyoto, there will be a cost to bear, but it won't be much. We could have blown past Kyoto without economic ruin. Indeed, hitting Kyoto targets could have made us *wealthier*, if we had done it intelligently.

It is equally untrue that the costs associated with the deeper, longer-term cuts required to stabilize the atmosphere would bring about our economic ruin. Danish political scientist and disgraced academic Bjorn Lomborg has become the darling of all those seeking comfort from inaction on climate disruption. Lomborg is no economist, and he's certainly no scientist. Instead, he's a statistician, whose cold, hard economic analysis generates alarmingly precise numbers that show that climate disruption is "a bit of a problem," but that the best thing for us to do is . . . to not worry too much and simply wait.

Lucky us. According to Lomborg, all we have to do is invest a few crumbs into R&D and wait for a technological magic bullet. All the heavy lifting can be borne by someone else (someone not yet born, conveniently enough). Lomborg is wrong.

If we take the cost calculations of the most pessimistic and vociferous defenders of inaction at face value and without poking holes in their models or questioning their assumptions, what is their estimation of the cost of aggressive cuts in carbon? What is the nature of the impending economic ruin that holds us hopelessly paralyzed? It might cost us three years of growth between now and 2050. In other words, we need to wait until 2053 to be more than twice as rich as we are now, instead of 2050. Another way of putting it: without measuring any benefits, it'll cost us a coffee and donut per person per week.

Three years of unrealized growth over the next half-century — that is the sacrifice our political and economic elite tell us we are unwilling to pay. Instead of having the facts laid out in language we can understand and that enables us to make informed choices at the ballot box, we are warned by our leaders of near-apocalyptic ruin. Of course we're not voting for strong solutions — we're told we can't do it!

Not all economists take this view, of course. Many quietly and clearly advocate for some sort of carbon pricing signal as a sensible, market-based solution. But the idea of economic cataclysm gets amplified in public debate when the likes of Abbott and Harper use climate pricing as a wedge issue to give cover for their climate quietism. Indeed, when it was pointed out to Abbott that the majority of Australian economists actually back the carbon tax, he responded by trying to bat away that expert opinion as one would an annoying mosquito: "Most Australian economists think that a carbon tax or emissions trading scheme is the way to go. . . . Maybe that's a comment on the quality of our economists rather than on the merits of the argument."[10]

Those defending the status quo have more than one hill to fight on, and economics is what you hit when you get past science. While public economic discussion is currently dominated by the views of those who have a somewhat hysterical view of the costs we might incur and be able to bear, there is no reason to think that this is for any reason other than political expediency. However, the transition to the language of economics from that of science does mean we've moved on from dismissing the problem outright to at least speaking about potential solutions.

Glenn Beck as Commie Stooge

Another act in the theater of the absurd is the painful irony of watching the U.S. let an opportunity to lead the new economy slip away, the way it has with the digital revolution. While the U.S. waffles on climate disruption, the command-and-control economy of China is winning the fight for the biggest economic score of the twenty-first century. And China is playing the U.S. for a fool. Worse, it is the likes of Glenn Beck and Jim Inhofe, those most bellicose defenders of the free market, who are playing the role of China's unwitting stooge.

The U.S. has walked into a trap. To placate the far right, the U.S. has stated it will not commit to hard emissions targets as long as China refuses to do the same. The Chinese know this, and they use it to their advantage. By refusing to agree to binding targets, China is ensuring the U.S. remains politically hamstrung, unable to pass the carbon legislation that would unleash American industrial might.

Beck and Inhofe's far right, paranoid, climate-scam rhetoric may make great populist theater, but the net result is that China is keeping the U.S. on the sidelines while it captures the lucrative clean-energy market—and all the jobs and investment that go with it. Beck and Gingrich are hurting America's economic recovery.

Make no mistake, China may be recalcitrant in climate negotiations, but the Chinese leadership fully understands the threat of climate disruption. Like most of the world, they know a global economic transformation to low-carbon energy is now inevitable. The centrally controlled Chinese economy and its national bank are pouring more investment and state support into clean energy than any other country by far. This includes not just money for energy projects but lots of cheap capital to build the factories that make solar panels and wind turbines. That's why China now supplies most of the world's massive and growing demand for solar equipment. While America bickers, China is quickly and strategically becoming a clean-energy industrial powerhouse.

While Inhofe is raging against the climate scam, China is pouring low-cost loans into state-owned wind farms, leveraging their demand for turbines to give Chinese manufacturers global advantages in scale. While Beck is railing against a perceived communist-driven conspiracy, China is establishing a massive lead in low-cost solar production. While they both ridicule our scientific elite, providing comfort to those who would ignore their warnings, another American solar company is closing its local factories and moving to China.

Beck's far-right ignorance of climate science and paranoid populist rhetoric about Marxist conspiracies beggars our democratic free market. A price on carbon is not a left-wing conspiracy to control the world. It is the best tool in our arsenal to unleash the might of our industry, capital, and entrepreneurs against the very real threat of catastrophic climate disruption.

In sapping our political will to act, Inhofe is playing right into China's hands. By the time China plays nice and agrees to hard targets, it'll own the clean-energy sector. So when the U.S. finally passes its own legislation, it will meet those targets by buying low-cost Chinese technology. We'll go from buying oil from the Middle East to buying clean energy from China.

Just Get the Bad Guys Out of the Way

Those at the other end of the spectrum, those agitating for a response to climate disruption, often imply that if we just got all those "bad guys" out of the way—the big oil and gas companies, in particular—the carbon problem would naturally resolve itself. If we stop the obstructionism of global energy companies and unleash the full power of our markets and creative class, then we'd solve this problem from the bottom up. The blogosphere is rife with optimistic visions that would come into play once we obliterate bad ol' Big Oil. Cars might already be electric or our fuel derived from biomass. Give the power to the people and the power will come from the people—a solar panel on every roof, an electric car in every garage, that sort of thing.

It's naive to think that we're going to rebuild a massive and robust energy structure—one that represents the work of several generations and trillions of dollars of capital—in this fashion. It may be absurd to deny climate disruption in the first place, but it is equally absurd (if somewhat more forgivable) to think the problem will go away along with the Exxons of the world. The job is much more difficult than such optimists would have it. Indeed, we can't solve climate disruption at all without the active cooperation of the (reformed) "bad guys." Exxon is not going to be replaced; it must be forced to evolve.

The alternative to centralized, corporate-dominated energy production is likened to a kind of distributed intelligence. Modern crowd-sourcing techniques, such as Wikipedia, and open-source collaborative software projects, like Linux, point to the highly leveraged power of individual creativity when given a minimal collaborative structure. It worked for software, why not for energy? Collaborative exercises like MIT's CoLab, which "seeks to harness the collective intelligence of people around the world to address climate change,"[11] do indeed accelerate the formation and adoption of great ideas, provide policy alternatives, and promote social action and awareness.

Don Tapscott, author of many books about the networked economic model, is as excited as anyone about the masses' potential contribution to solving the climate crisis: "Around the world there are already hundreds, and probably thousands, of collaborations occurring; everyone from scientists to school children are mobilizing to do something about carbon emissions. And the most forward-looking political leaders recognize that amplifying these grassroots energies could be our best short-term hope for meaningful action,"[12] Tapscott writes. I agree. That sort of shared thinking might be the best we've got on the table so far, but by itself it will never amount to much. Our optimism for collaborative, grassroots action on the energy front is misplaced if it is not accompanied by policy that directly engages the existing energy giants.

It's true that energy systems will become much more distributed, with local wind farms, rooftop solar panels, and electric car batteries that act as storage for clean energy. Home- and factory-level intelligence will be part of an emerging Energy Internet—a complex dance of energy across the continent that balances production, distribution, and use. All these characteristics are shared with the open-source, collaborative model at a very high level.

But building low-carbon energy systems, even with lots of distributed solar panels and community-owned wind farms, requires infrastructure and finance on such a massive scale that we're not going to build it without the cooperation of the big players. With small exceptions, the choices we have for our energy consumption patterns will remain something that is presented *to* us, not created *by* us.

Even if you want to run your car on fuel derived from biomass, you can't do so without a national-scale infrastructure to supply it. Only a continent-wide grid could deliver electricity 24/7 at anywhere near current consumption levels. Solar panels on roofs will lower demand, but we'll all still be connected to a grid fed by large power plants. We might replace coal with heat from the ground, but we are not going to do away with large energy companies. Individual autonomy in energy production and consumption will be a niche market at best—an interesting anomaly, like eggs from backyard chicken coops or fresh-from-the-cow milk.

De-carboning the economy is the largest infrastructure project in human history. We are talking about replacing the largest and most complex piece of machinery the world has ever seen, one whose output permeates every commercial transaction, from agricultural exports to toys and from our morning cup of coffee to our annual vacation. Cars, lights, iPhones, water supplies, the tires on our bikes, the wires in our walls—all of this stuff sits on a huge web of energy that flows from country to country, pump to tank, power plant to wall socket. Since we don't have the luxury of designing the continent's energy infrastructure from scratch, we need to replace all this while keeping the lights on and the factories humming. All that work comes with huge scientific, engineering, and financial challenges. The amount of capital we need to deploy is in the trillions of dollars.

To get it done, we'll need the participation of global energy giants, engineering firms, and the banks that back their projects. Pension funds are the only institutions with a timeframe that matches climate damage, and they need to come on board—perhaps with arguments about fiduciary duty to their beneficiaries or with good old-fashioned long-term returns on investment (or, better still, both). We'll need the endorsement of the majority of our professional and management classes, who must be convinced that they can be a part of the solution with minimal economic and reputational risk. We'll need the expertise, balance sheets, and industrial might of companies like BP if we're to drill for heat rather than oil.

Sometimes we mistake legitimate condemnation of the often obstructionist and recalcitrant stance of the energy giants and other established industrial players for a belief that we don't need them to be a part of the solution. Getting vested interests out of the way so that good policy and public debate can move forward doesn't mean that we can implement the resulting solutions without them. Like it or not, environmentalists must evolve to see that the big energy companies will be as much a part of future solutions as they are a part of the current problem.

The Conservative Paradox

What if there emerged a problem that *required* centralized economic control? A problem that required us to think in ways normally associated with (shudder) socialism? If such a problem were even thinkable, let alone possible, it would look an awful lot like the climate crisis. It should be no wonder there are libertarian climate quietists on Fox each night saying, "Thank God it's not real, praise be to Greenspan! Because if it *is* real, we've got no tools in our arsenal to deal with it. Um. Uh-oh . . . What if . . . Nah, can't be true!"

Views on climate disruption are strongly correlated to political affiliation, particularly in the United States and to a lesser extent in Canada. Progressives tend to see climate disruption as a threat of some urgency, while conservatives, particularly the modern brand of neocons, tend to downplay or even deny the threat. Neocons have been forced to take their stance because climate disruption reveals a very difficult paradox at the heart of modern conservatism that applies equally well in both Canada and in the United States: business as usual will no longer preserve the status quo.

Until conservatives resolve this paradox, it will be very difficult for them to be actively engaged on the issue. The problem is that hard-line, ideological neocons who cannot prioritize their theoretical commitments in response to empirical data are not able to grapple with climate disruption.

Whatever more radical form conservatism may take—be it Bush-era neoconservatism, libertarianism, or the more recent Tea Party movement—two ideas lie at its philosophical core. The first is the notion that we make radical changes to our social, political, or economic structures at our peril. The conservative adopts a cautious and incrementalist approach in order to preserve what we have: life's pretty good, so let's not make any fast moves. This approach has often served us well, acting as a counterweight to all sorts of social and economic experiments—a steadying hand on the tiller. The second is the power and sanctity of the market. Trusting the market's invisible hand, leaving it as clear of interference as possible, is part and parcel of conservative thinking.

But we are now faced with the paradox of having to make radical changes to our energy economy if we have any hope of preserving something close to our current way of life. Business as usual will bring significant and unpleasant changes to our climate, which in turn will wreak economic havoc. Hence, radical change will come unless we make radical changes. Worse, making those changes requires more than mere fiddling with the market. To unleash its power, the market will need to be shaped, even tamed. If there were ever a problem that could challenge the core of conservative thought, it's climate disruption.

There are three ways for conservatives to resolve this conflict (in philosophy, this is called a *trilemma*): find a solution that maintains all conservative values, compromise on one or more of those values while emphasizing another, or ignore the problem. I do not believe it's possible to keep all conservative values in play while seeking a solution in good faith. So option one is out. Number two is an option for flexible (or moderate) conservatives. Option three is the only out for a brittle, unbendingly ideological neocon, which, presumably, is why they work so hard to perpetuate scepticism.

Repositioning conservative values means emphasizing one value at the expense of another. For example, conservatives might ditch adherence to incrementalism while fully committing to the most important status quo of all: keeping our basic economic and civic structures intact. Conservatives might then choose to soften their free-market rigidity. In return, by accepting the need for market signals, they can bask in the acknowledgment of the marketplace as the most powerful tool we have. Conservatives love the market, so this should be a no-brainer—their darling saves the world. The pain of accepting market interference should not be too bitter a pill to swallow.

Some say market interference, while others might say market-based solutions. Pricing carbon as an externality is Economics 101. By that, I mean pricing externalities is not controversial and does not run counter to any core conservative values or even basic market theory (other than contradicting a circular argument that relies on the premise that free markets are good). It can be defended on any number of grounds. This is clearly a move that I, and other market-oriented entrepreneurs, endorse. That endorsement is politically neutral; it is simply an effective response to an urgent problem, nothing more.

But so far a different voice has dominated the conservative reaction to the paradox on climate disruption: with notable exceptions, modern conservatives have had to play to a neocon base, and they decided to simply deny or downplay the problem. This makes things very difficult, since — as this book will make clear — public opinion on climate disruption is particularly malleable. There is some irony in the fact that, until it evolves, the conservative voice is a significant hurdle to the preservation of the status quo. Conservatives are star players in the theater of the absurd, and there is no reason this need continue.

Once conservatives become less rigid and accept that a rebalancing of core commitments is both possible and necessary, their choice is reduced to a very manageable dilemma. Dilemmas are often compared to the sort of choice Homer's Odysseus faced. He was tasked with charting his ship on a very narrow course, avoiding the rocks of Scylla on one side and the whirlpool of Charybdis on the other. The modern conservative paradox reveals a similarly narrow set of options: on one side are rocks that might destroy incrementalism; on the other is a whirlpool that might sink the free market. The way through is to embrace action in the form of market-based solutions. Give up incrementalism in favor of the power of the market and a broader status quo.

WAKING THE FROG: EXPLAINING THE ABSURDITY

Some of the dynamics driving the public debate are not hard to pick out and aren't controversial. Public relations firms and think tanks hired by fossil-fuel-dependent industries have sown seeds of doubt about the science behind climate disruption. A cacophony of voices, representing the opinions of just about anyone who cares to speak on the matter, is drowning out the measured tone of scientific organizations dedicated to the study of geophysics and atmospheric chemistry. The conservative political movement, which is dominated by neoconservatives and their mouthpieces, has pushed back—hard—against anything that threatens to intrude upon the free and unfettered operation of the market and those with vested interests in it.

These voices and ideas land on a public that is eager—indeed, psychologically predisposed—to believe things can continue as they are and that the warnings of calamity must be false. The resulting public scepticism about the need to take action then informs the political debate, and on it goes.

Meanwhile, the scale of the required response—the sheer mass of energy infrastructure that must be replaced—has become dauntingly clear. To make matters worse, we won't reap any benefits (at least on the temperature front) from concerted efforts to change that infrastructure for many decades. Any heavy lifting we do now will result in only very small changes, decades out, in the centuries-long temperature arc we're riding. So even those who have seen the dangers ahead and are convinced action must be taken often find themselves rooted to the spot, bewildered by the amount of work to do and discouraged by the chasm of time before we'll see results. Motivation to act—like we saw back in late 1980s, when Margaret Thatcher made her landmark address to the United Nations—can easily turn into fatalistic acceptance.

All of this poisons the political atmosphere and paralyzes action, but it is possible to look past these surface dynamics to reveal what I see as a deeper structure. Far from being a cacophony of voices and opinions, I think there are clear delineations in this debate that expose a few deep, underlying faults. Our socioeconomic machine is powered by psychological dispositions, ways of understanding the world, and corporate, legal, and political structures. Revealing the workings of that machine, even in very simple ways, might better enable us to break out of our paralysis.

Suggested Reading

Gwynne Dyer, *Climate Wars: The Fight for Survival as the World Overheats* (2008)

Clive Hamilton, *Requiem for a Species: Why We Resist the Truth About Climate Change* (2010)

Mike Hulme, *Why We Disagree About Climate Change: Understanding Controversy, Inaction and Opportunity* (2009)

James Lovelock, *The Vanishing Face of Gaia: A Final Warning* (2009)

George Monbiot, *Heat: How to Stop the Planet Burning* (2006)

CHAPTER 2

The Siren Song of Denial

"The path of least resistance is the path of least change."
— John-Paul Tamblyn, yogi

"Ideology is a conceptual framework, it's the way people deal with reality.
Everyone has one. You have to. To exist, you need an ideology."
— Alan Greenspan, U.S. economist

In Greek mythology, sirens were seductive women who sang irresistibly beautiful songs to lure sailors. Unsuspecting seafarers, mesmerized by the music of these femmes fatales, were enticed to sail ever closer to the rocky coasts of the sirens' islands. Listening to the sirens' songs ensured a horrible fate, as the sailors' boats crashed upon the rocks. Climate denial is a kind of siren song. It's seductive to believe climate disruption might not be true or be as bad as it's made out to be, but following the siren song of denial takes us into very dangerous waters.

Nobody in their right mind wants climate disruption to be true. Only a sociopath would take any pleasure, schadenfreude, in being right on this one. Even people who work hard to seek the truth about climate disruption have trouble believing it's quite as bad as the evidence would have it. That's because it's hard to believe and easy to deny. Our minds will play all sorts of tricks to keep us from absorbing that truth. It's perfectly natural to heed the siren song of denial, which makes it even more dangerous.

Denial comes in many forms. You don't have to think that climate disruption is a load of bunk to be in denial. Some accept it's real but deny the consequences. Some ignore it altogether. Lots of people sit somewhere in between: the suv driver who can't admit he's contributing to the problem, the captain of industry who wants to keep turning a profit the same old way, an anxious mother who turns her thoughts to something less worrisome. The most slippery form of denial is to acknowledge the problem but dismiss the consequences. Politicians who admit climate disruption is a threat yet speak of allowing carbon counts to rise to 550 parts per million (ppm) because it's politically expedient are in denial. We are all in some degree of denial. It's natural because climate disruption is scary.

Marketers have known for decades that we're not rational creatures. A good advertisement doesn't focus on facts but gets us to respond emotionally. Psychologists and cognitive scientists have mountains of evidence that thought is rarely logical, even without the influence of a good marketing campaign. All sorts of stuff—emotions, personal history, fear, desire—get in the way of making a logical decision. *Star Trek* fans have always known we have little in common with the Vulcans. Mostly, we humans believe what we want to believe.

Normally that's not a problem. As long as we know how to pay the rent, feed our kids, drive our cars, and hold down a job, we're pretty much free to believe whatever we want. Perhaps free trade will bring more jobs, perhaps it won't. Maybe buying that new car really will make me happy. Blowing on the dice in Vegas will make them come up seven. Believing what we want sometimes comes at a price, but it's one we can normally afford. But some beliefs matter a great deal and affect us all. It matters whether or not we believe DDT causes cancer because we need the political will to ban it. It matters if a politician tells the truth about the reasons for declaring war. And it matters that we believe our carbon emissions are warming the planet.

When Galileo discovered that the Earth orbited the sun and not the other way around, his idea was met with disbelief and anger. He was threatened with excommunication from the Church. His discovery ran counter to accepted wisdom. It ran head-on into other deeply held beliefs. It was important back then to believe humans were at the center of the universe. All sorts of religious, cultural, spiritual, and emotional systems depended on it. A threat to those beliefs caused negative emotions, like fear and anger. Brains soaked in medieval culture would not easily let in this new belief. Eventually, culture changed, and we now know Galileo was ahead of his time.

When belief in climate disruption comes knocking at your mind's door, it runs up against all sorts of cognitive machinery working overtime to keep it out. It bumps into other deeply held, but incompatible, beliefs: the future is better than the past, economies grow forever, technology and markets solve all problems, humans can't really bake the whole planet. Add to the mix an explicit campaign of disinformation and a growing awareness of how hard it's going to be to stop, and it's no wonder we're all in some kind of denial! But avoiding difficult beliefs is no help. We teach our children to face up to their fears. We need to do the same ourselves.

In this chapter I explore *why* it's hard to really believe in climate disruption and its consequences and so easy to rest blissfully in denial. I am no different: the cognitive and cultural machinery I discuss is common to all of us. In understanding how our minds work, I hope we can learn better ways to communicate about the risks we face: by understanding the allure of denial's siren song, we might avoid its trap.

Duplicitous Denial

Denial is a real psychological phenomenon and has long been studied in relation to damaging behavior like spousal abuse and alcoholism. A person in denial is unable to acknowledge a state of affairs that seems obvious to an outside observer. Someone denies they're an alcoholic even though their drinking costs them their job. A woman denies being in a bad marriage even though she goes to bed with a black eye. In clinical cases it's often construed as a means of self-preservation. In cases of sexual abuse, for example, an essential sense of self is protected when one pushes away an unpleasant truth from one's past. You don't have to actually deny something to be in denial: ignoring something relevant is denial, too. Freud made hay with this stuff.

For simplicity, I divide climate denial into two types: active and passive. People who openly claim that climate disruption is not true—the planet is not warming or we're not the cause—are in active denial. It's a cultural niche dominated by extremist "sceptics" who ensure their view takes up a lot of Internet, political, and media space. Passive denial is the more subtle ability to ignore climate disruption. Lots of people do so, sometimes even while acknowledging it's real. Active denial is loud; passive denial is quiet. Active denial is aggressive and up-front; passive denial can be sneaky.

I think some loud active deniers, and those who fund them, are disingenuous.[13] They don't believe what they say. They are being strategic, protecting profits, increasing a voter base, or grabbing media ratings. Most active and passive deniers are perfectly reasonable people getting on with their lives. Denial is not necessarily blameworthy. It's a natural response to a frightening existential threat, but we need to face that threat, so *any* form of denial is unhelpful. But a few really loud active deniers do a lot of damage by popularizing a series of myths that make it easier for the rest of us to ignore the greatest threat of our time.

Conversations with Brian

One of my first encounters with an extreme sceptic happened by chance in a prolonged exchange with an old high school friend. Let's call him Brian.[14] By "extreme sceptic" I mean someone who vehemently and angrily denies that anything like climate disruption is real, no matter what evidence you put in front of them. I was astonished at the degree and obviousness of the self-contradictions Brian was able to maintain without any seeming awareness of doing so. When I pointed them out, he would abruptly change the topic, accuse me of conspiracy mongering, or get so angry I had to let it go. I have since learned to avoid these sorts of exchanges.

In Brian I encountered an old-fashioned, Freudian kind of near-clinical denial. I was fascinated enough by our exchange to document it. It's instructive because it includes most of the popular denial touchstones that were floating around at the time. Here is that exchange, edited for brevity:[15]

Brian: There's no evidence for global warming. There hasn't been any warming since 1995.

Tom: That's statistically naive. It's like picking out a few down days in a long-term bull run on the stock market and saying there's no bull market. You can always pick short-term bits counter to the long-term trend. That's just noise. The long-term trend is up.

Brian: Well, the Earth's warmed before. This is just part of another natural cycle. We adapted then and we'll adapt now.

Tom: Either it's warming or it's not. Can we agree it's warming?

Brian: Crisis over. Next on the elitist agenda is biodiversity, like we're causing a bunch of extinctions. It's just another excuse for a carbon tax.

Tom: I wasn't talking about biodiversity. First you said it wasn't warming, then a second ago you said it's warming but natural. I just want to know if we agree it's warming or not.

Brian: Look, there is no scientific consensus. The IPCC is politicized,[16] and Climategate shows they lied about humans being the cause of global warming.[17] It's a bunch of Green Nazis who are responsible for public fear of global warming, not empirical evidence!

Tom: Whoa, wait a second. Climategate showed scientists are human and make mistakes, but it doesn't change the basic facts. And the data under question came from multiple sources. And where on earth do you get "the IPCC lied about carbon dioxide being the cause of warming"? Where does that come from? Back to basics: is the Earth warming, or isn't it?

Brian: Science requires scepticism, that's what I'm doing. If a scientist isn't sceptical then they're just a politician.

Tom: What are you saying—that climate scientists aren't real scientists because they take a position? And you are that scientist? [Brian has never studied science.] You haven't answered my first question: Do you agree it's getting warmer?

Brian: This year was the coldest and snowiest December in Europe since records have been kept! More and more people are waking up to the fact that global warming is not about science but about control.

Tom: We're back to statistics and noise. Pointing out a cold December in Europe is statistically irrelevant. Do you think you have the expertise to make better judgments about long-term temperature trends than NASA? Seriously?

Brian: There's still tons of polar bears. It's all bullshit.

Tom: I wasn't talking about polar bears. You don't believe NASA when they say it's getting warmer? Or is it getting warmer because of a natural cycle?

Brian: Pointing out Europe is getting colder shows you global warming alarmists can't predict your way out of a paper bag. Besides, dude, only four parts of ten thousand in the atmosphere is carbon! It's irrelevant.

Tom: Weather isn't climate. We can't predict the weather next July, but I can tell you with certainty it will be hotter in July 2100 than December of that year. A cold December in Europe is irrelevant — just like the down days in a bull market. About the tiny amounts — a few parts per trillion of some poisons can kill you. So tiny is relative. Your point doesn't make sense.

Brian: Besides, plants need carbon dioxide to flourish! Try some common sense!

Tom: Sure plants need carbon dioxide. Who denies that? But didn't you just tell me there was too little of it to make a difference? Besides, our bodies need salt but too much of that can kill us. You're not making sense. I still want to know if you think it's getting hotter — we haven't even settled that yet!

Brian: This whole thing is an invented crisis to galvanize a global government — same as the economic crisis.

Tom: The whole thing was invented?! You know it takes thousands of scientists, over many decades and many countries, to "invent" climate change, right? But I want to keep on track. I still can't tell if you're saying there's no warming or it's a natural cycle . . .

Brian: It's the sun, moron . . .

And on we go, round and round. I'm no professional psychologist, but surely this level of denial is clinical.

Brian represents a tiny fraction of the public that engages in this kind of open self-contradiction and anger. But Brian's ideas originate from the outpouring of a small portion of the blogosphere: unscientific, non-peer reviewed websites like ClimateDepot.com and WattsUpwithThat.com that act as the wellspring of a well-oiled denial machine. The talking points are repeated by hundreds of other bloggers, like a giant bullhorn. They are absorbed and repeated by millions of people, from your neighbor to Margaret Wente to Fox News. They are treated as facts in the great climate debate. The Internet becomes a giant echo chamber that feeds into popular culture the kind of utter nonsense Brian has internalized.

Brian personifies what happens when you try to absorb, all at once, all the climate crap that floats around on the Internet. He's a complicated, contradictory, and very angry person. Once the uncomfortable dance of denial begins, we can dismiss him as a bit out there, just as we can a lot of angry and often incoherent public figures like Rush Limbaugh and Glenn Beck. But these people are persuasive professionals. Only when you have a chance to push them as hard as I did Brian do their arguments begin to sound unbalanced.[18]

The ideas that Brian has internalized—it's a natural cycle, carbon dioxide is a harmless plant food, there is not really any warming—have all permeated the collective consciousness of the general public to some extent. Not everyone believes all of them, of course, but as you chat at dinner parties or over the backyard fence, I'm sure you'll find there are seeds of these ideas in almost everyone.

Reasonable People

Taken all at once, these ideas generate a nutty dance of avoidance. But any *one* of these ideas can be a reason to deny what's happening for someone who doesn't want to believe the worst. Some provide reason for active denial: temperatures are not really rising, carbon dioxide is only a harmless plant food, there's lots of disagreement in the scientific world, etc. Some provide reason for passive denial: we can adapt, it won't be so bad, this warming is just part of a natural cycle, etc. The Internet's echo chamber has lots of tasty bait for anyone who prefers not to worry.

Lots of perfectly rational, smart people take these positions. An educated professor friend of my father's thinks it's a natural cycle. If you push him on it you'll find out he doesn't know Malinkovitch cycles from a rubber shoe.[19] But if he wants to be in denial about climate disruption, it's easy. All he has to do is swallow some of the bait and not think too much. Getting past the bait takes work, swallowing it doesn't.

We can also accept that climate disruption is both real and dangerous yet still turn a blind eye. You don't need a reason to ignore it, you just do. Recent surveys in the United States show most Americans are now convinced climate disruption is happening: fully forty percent believe the consequences will be severe in our lifetime.[20] No denial there, but we still manage to push that awareness into the background. Even after Hurricane Katrina hit in 2005, Gallup's headline for its annual environment poll read "Americans Still Not Concerned About Global Warming." As an indicator of the effects of failed civic infrastructure in the presence of increased extremes of flooding, Katrina was a harbinger.

Sociologist Kari Marie Norgaard calls this "climate change as background noise,"[21] which she characterizes as "the possibility of climate change [being] both deeply disturbing and almost completely submerged, simultaneously unimaginable and common knowledge."[22] Turning a blind eye to something disturbing is an old trick of the mind. We want to control negative emotions, and according to psychologist Morris Rosenberg, "the main way of controlling one's emotions is to exert control over one's thoughts."[23] So we turn away from the source of those emotions.

Historical psychologist Robert Lifton uses the term *psychic numbing* to describe the way people react to the threat of nuclear weapons. His research started with Hiroshima survivors. They were unable to respond rationally to their surroundings. No surprise there. More importantly, his work found that the existential but abstract threat of nuclear terror that we lived through during the Cold War led to a similar "age of numbing."[24] I see no reason why our inability to integrate climate disruption into our lives, and react rationally to this profound and existential threat, is fundamentally different.

Industries and institutions can be active or passive deniers, just like individuals. The oil patch plays both roles. Exxon Mobil and the Koch brothers are champions in active denial, funding spin-machines like the Heartland Institute,[25] which was one of the first critics to push the "carbon dioxide as harmless plant food" line. Deliberate passive denial is the oil patch quietly going about its business, ignoring climate disruption while being careful not to actively deny it. In a debate I had with David Collyer, head of the Canadian Association of Petroleum Producers (CAPP), he agreed we need a policy on carbon but declared CAPP was formally "agnostic" on climate disruption. That makes no sense. What's the point of a carbon policy if climate disruption is not real? Mr. Collyer is in denial.

There are lots of reasons for passive denial. It's not unreasonable. We may feel powerless or overwhelmed. We might want to avoid feelings of guilt. We are afraid of what it means for our children. But there is no excuse for civic leaders to engage in active denial. It's their responsibility to know better.

Imagine we're on a spacecraft, and the mechanic has discovered a leak in the airlock. His team can fix it, he says, but everyone will have to cooperate by helping to pay for a new airlock. No problem—a bit tough, but we can do it. Suddenly, a cheapskate, loudmouth crewmember starts telling everyone it's a waste of money. He's got a homemade pamphlet that says we don't need that much air. Another, beefier, crewmember tries to stop the mechanic from getting to the airlock. These two loudmouths represent civic leaders who play up denial. They are a danger to everyone.

If this strikes you as an exaggerated metaphor, I'd point out that a Pentagon report to President G.W. Bush in 2003 states that climate disruption is "a threat to global stability [which] vastly eclipses that of terrorism . . . [and it] should be elevated beyond a scientific debate to a U.S. national security concern."[26]

We've become quite scared of people blowing up buildings and planes. In relation to my earlier metaphor, terrorism is like one of the spaceship's toilets being plugged. We've got an awful lot of staff keeping the toilets open. We might want to get a bit tougher on those keeping us from fixing the airlock.

Many colleagues involved in clean-energy and climate advocacy quite naturally lament this state of affairs. As long as we remain in denial about the danger we face, it's impossible to galvanize action to solve the problem. One solution they point to is information. A common refrain is "If only we could educate people, they would realize the truth and take action!" This, unfortunately, is wishful thinking.

RATIONAL? WHO'S RATIONAL?!

I grew up in an academic household. My father was a professor of molecular biology, and my mother was the first female engineering graduate from Yale. As a young man, I shared their scientific worldview that people are rational and public dialogue settles on the truth. Oh, most of us might not be perfectly rational and public dialogue might have to go on for some time before settling on what is good and right, but if you put the right information out there, we'd eventually get there.

My parents were wrong.

Science, technology, and rational thought are linked. Science sits on a foundation of rational thinking and empirical evidence. Technology is science made practical by engineers. It might seem reasonable to assume our society is rational, given how much technology and science inform and enable modern life, but if being rational means forming beliefs based on information, logic, and empirical evidence, then it's naive to think we're rational. Psychologists and philosophers have spent a lot of effort showing rationality is a kind of illusion. Before we get to that, there are more obvious ways we're irrational.

For one thing, lots of us are religious. Religion is an explicitly irrational system of beliefs. That's why it's called *faith*. Each faith has its own favorite beliefs. The Book of Mormon, said to be a transcription of a set of gold tablets shown to Joseph Smith by an angel, talks of American Indians as direct descendants of Israelites and Jesus visiting North America. There is no evidence for any of these claims, including the existence of the gold tablets. Other religions hold beliefs that run counter to rational evidence: the Earth is only a few thousand years old, women get pregnant without sexual intercourse, and so on.

I say these things not to condemn religion — it is what it is.[27] The point is that people exercise their freedom to form beliefs whatever way they choose, rational or not. We form beliefs for all sorts of reasons: our faith, our friends, it makes us feel good, it gives us a code to live by, or often just because we grew up that way.

Late one night, my friend Toby Heaps, the founder of *Corporate Knights* magazine, and I were lamenting the difficulty experts have had convincing the public to take climate disruption seriously. Groups like NASA, the National Academy of Sciences and the Royal Society have been unable to galvanize public concern, even though both are filled with Nobel Prize winners and some of the smartest minds on the planet. What would it take? So we performed an experiment. We asked

Toronto cab drivers (a diverse and notoriously well-educated crowd): "Who could convince you climate disruption is real?" Answers varied, but it sure wasn't the scientists studying the stuff. Most of them picked their family, a religious leader, or a celebrity actor.

Belief often comes down to trust. Trust can be based on historical or social connection, emotion, charisma, celebrity, or almost anything. Charisma is particularly effective. Some people trust their friends, some their priests, and some the august traditions of empirical science and the pronouncements of the Royal Society or the National Academy of Sciences. To each their own, at least here in the free world.

Even When We Try ...

Turns out we're not all that rational, even when we try. Psychological and cognitive science literature is jammed with examples of our imperfect rationality. Doesn't matter that we're focused on the task of being logical. We're terrible at probabilities, don't remember the past accurately, our judgments are skewed depending on how a problem is presented, and answers are affected by all kinds subconscious suggestions. Here are just two short examples.[28]

Recall that Italy played France in the 2006 World Cup final. It was an exciting game and included that strange moment when French captain Zinedine Zidane head-butted Italian player Marco Materazzi. The game can have only one winner, so these two sentences are logically identical: "Italy won" and "France lost." This means that if we were perfectly rational, they would mean the same thing to us, but they do not. The first automatically brings up positive thoughts: winners, victory, celebration, and so on. It might also bring up thoughts about your own personal history with Italy, a vacation, good wine, and so on.

The second brings up negative thoughts: loss, regret, pain, and so on, as well as some personal thoughts about France.

The statements are logically identical but very different to us humans. We'll discover why in the next section, but the short answer is our minds do not process information logically. Our mind automatically associates one thought with another. Thus, each of us will have different reactions. There's nothing we can do to stop those associations. We can't prevent them from affecting our reaction, our next thought, and so on. The words we choose matter, the way we say them matters, and the context into which we put them matters. This may sound like common sense, but it has profound implications for efforts to combat climate denial. This very simple example holds lessons for those of us trying to move the needle on public engagement.

Here's another one.[29] What follow are descriptions of two people. What do you think of Alan and Ben?

Alan: intelligent, industrious, impulsive, critical, stubborn, envious
Ben: envious, stubborn, critical, impulsive, industrious, intelligent

Most people favor Alan over Ben, even though their traits are exactly the same (just presented in reverse order). We place more weight on the first trait. More interesting is how the initial traits affect the way we interpret those that follow. Stubbornness in an intelligent person might be regarded as admirable. They stick to their guns. But intelligence in an envious person might be seen as evil. In addition to automatically associating different thoughts, our brains work to interpret what's in front of us — and what's in front of us usually contains a lot of ambiguity. Alan and Ben are not five words, they're people. We fill in the gaps.

So guess what? We're not computers! This is no surprise. Still, ever since Descartes declared, "I think, therefore I am!" much of Western culture assumed some (if not all) of our thinking is abstract, logical, rational. That assumption underpins most of modern economics, as we'll see. According to linguist, philosopher, and cognitive scientist George Lakoff, "most of us have inherited a theory of mind dating back at least to the Enlightenment, namely, that reason is conscious, literal, logical, unemotional, disembodied, universal, and functions to serve our interests. This theory of human reason has been shown to be false in every particular, but it persists."[30] Lakoff makes the point in his recent book *The Political Mind* that this has profound implications for how political battles are to be fought. It's equally true for engaging the public on climate disruption.

Rational Enough?

Cognitive scientists can invent tricky problems that show we're not perfectly rational. So what? Our thinking might not be perfectly rational, but it's pretty good. We can balance the household budget, figure out priorities for the day, raise our kids to be decent people, and get on with life. We are *pragmatically* rational. We make sense of the world in a "good enough to get by" sort of way. We have common sense. Surely we're rational enough! Haven't we humans been dealing with threats in a pragmatic way for aeons? And we have enough common sense to place our trust in the right experts when it comes to matters of great importance, right?

Our health and that of our loved ones is a good test: when we're deathly sick we place our trust in doctors; when our tooth hurts, we go to the dentist; when our dog gets hit by a car, we take it to the veterinarian.

We wouldn't get on a plane that wasn't flown by a certified pilot or drive a car that wasn't designed by professional engineers. Even people whose religious beliefs forbid modern medicine go against those beliefs and put their faith in science-based medicine when the going gets tough (holdouts exist, but they are rare). We trust the right experts in matters of personal safety. Those experts are members of professions that are based on science and rationality. Science and rationality is foundational to the Western world, and we turn to it when it counts. Isn't that enough?

Yet that's precisely what we're *not* doing for climate disruption—even though our well-being and that of our children is at stake. We are not heeding the call of the National Academies of Science. We are turning away from the warnings of the International Energy Agency. We often choose to trust a good-looking weathercaster on Fox News and slick, oil-funded advocacy groups over real experts when it comes to this threat. Why?

Environmental groups tend to think there's an information gap. If only people knew the facts, it would provoke awareness. Realizing we're not rational should put an end to that notion. Indeed, sometimes the more we know about climate disruption, the less likely we are to act: "in sharp contrast with the knowledge-deficit hypothesis, respondents with higher levels of information about global warming show less concern."[31] To tackle this problem, we need to understand how we think.

HOW WE THINK

We may not be very rational, but humans are brilliantly intelligent, far more intelligent than an emotionless Vulcan or a computer will ever be. We have a different kind of intelligence, and it's far more useful to us than mere rationality. We do not live in the world of Gary Kasparov playing chess against IBM's Big Blue. We live in a highly complex world where we have to deal very quickly with all kinds of ambiguity. There are lots of things we can do that computers cannot.[32] We are *more* than rational, not less, but it comes at a cost. One of them is a predisposition to deny threats like climate disruption.

I've always been partial to the following example of human intelligence because it's so simple yet makes a profound point.[33] Can you make sense of the phrase "Stay close to me"? A perfectly rational computer-like person cannot. That's because rationality by itself is not enough to disambiguate the meaning. The phrase means one thing to an imaginary space captain in a story, another to two people creeping down a dark alley at night, another to two lovers, and yet another to a day trader setting up her trades for the day. And even with a defined context, each person will have a different meaning of "close" depending on their own emotional structure, their own history, and so on.

To make sense of that simple sentence we first have to pick out which other facts are relevant. We pick those out from our entire background of knowledge about the world. Are facts about muggers relevant? How about big cities, tables, criminals, poetry, or fear of stock market arbitrage? Then we use only those relevant facts to help make sense of it. How big is the set of facts that makes up our background knowledge? It's potentially infinite.[34] Picking out relevant facts to disambiguate meaning is a snap for us. It's impossible for a computer. Now try something like "Every time I've loved we happily shared melancholy."

There are lots of other things we do very easily that are difficult (or impossible) using a rules-only approach. Our unique human intelligence gives us the ability to instantly access a worldview. That's really what we mean by common sense. Our minds have instant access to a shared wisdom about the world and the people in it. We know how the world works, what's ordinary, and what is not. We deal with surprises in practical ways. In real time we apply those bits of common sense relevant to the task at hand, for our lives, and our interests at that moment. It's quite remarkable. The key point is that we don't use rules to make sense of language or the world around us, nor could we. There's just too much to compute. Rationality has limits, and we are not bound by these limits. Our minds use a different approach. While it has advantages, it also has hidden traps.

Two Minds

We really have two minds working together. One is very fast, automatic, and unconscious. The other is slow, methodical, and conscious. Nobel Prize–winning psychologist Daniel Kahneman calls them System 1 and System 2, while cognitive linguist George Lakoff calls them reflexive and reflective. In my own academic work, I have called them intuitive and symbolic. We all mean roughly the same thing. All the really interesting stuff happens in the intuitive processor. It feeds stuff up into our consciousness, where language and deliberative thought take place. It's part of what makes us different from computers.

The intuitive mind is an associative engine: its job is to automatically and immediately bring forward thoughts, ideas, and judgments relevantly associated with whatever it's being presented with. It can make judgments and recognize patterns. It's what allows us to immediately classify things we see without thinking. It does this without using rules because it is built out of a neural network, which is a completely different kind of processor than a typical computer.

Neural networks record impressions and make connections between impressions that are linked in real life. If you see a tiger and hear a roar, when you later see a tiger you will think "roar." A connection between ideas in a neural network is like a river that carves a path in the ground. At first the water can go anywhere. Over time it forms a groove. That groove is the association between one idea and another, part of a network of connections almost without end. "Tiger" gets linked to details like tiger, fur, and claw and to general ideas like Africa, zoo, and fear. There are uncountable numbers of real neural grooves in your mind that create a vast web of interconnected ideas, impressions, facts, and memories.

Our emotions play a crucial role, helping us rate the importance of ideas and experiences. They also add their own content. Our brains don't operate in the abstract, like the mythical brain in a vat.[35] Our brains are embodied and deeply integrated with the physical and emotional structure of our bodies. Our bodies provide perspective and a set of interests. Our unique perspective of "me," complete with its foundational emotional structure, enables an initial filtering of the mass of data around us into what's relevant. Without a body and emotions, we could not think the way we do; indeed we probably could not think at all.

We cannot control what happens inside our intuitive mind, but we do "program" it (in a way). We program it by living our lives. A musician trains himself to recognize and respond to musical cues. A doctor trains herself to recognize illness. A chess master trains himself to recognize strengths and weaknesses on the board, so he doesn't have to think about every possible move. These are well-studied phenomena, and the way neural networks do it is well understood, even if it sounds a bit mysterious.[36]

Over your lifetime, your intuitive processor effectively records everything you experience. It makes connections between ideas you put together through your experience of your life, lived in our shared world. Over time your intuitive mind reflects and instantiates (makes physically real) your view of the world, your common sense. That worldview informs each idea you form, the way you think, and your reaction to everything you experience. We need a worldview to think and to make sense of the world. U.S. economist Alan Greenspan was partly correct when he said, "Ideology is a conceptual framework, it's the way people deal with reality. Everyone has one. You have to. To exist, you need an ideology."[37] Greenspan's mistake is to confuse ideology with a worldview. Ideologies don't adapt to new information, but a worldview is constantly reshaped. (An ironic choice of words by the man nicknamed the "Oracle.")

Our intuitive mind is very good at instantly recognizing something or generating relevant similarities. Its core function is to associate one thing with another: that shape is a tree, that person is dangerous, I hear a B diminished chord, and so on. It enables us to make fast decisions about complex situations with incomplete information. We cope with open-ended ambiguity and complexity. The intuitive mind values coherence, consistency, and ease of access. It's what makes us *more* than rational. But all that associative power comes at a cost, since its strengths also bring weaknesses. It can simultaneously make us *less* than rational.

Recall Lakoff's list of the old way of viewing the rational mind: "conscious, literal, logical, unemotional, disembodied, universal, and functions to serve our interests." The intuitive mind is the opposite: it is unconscious, metaphoric, (somewhat) illogical, emotional, deeply affected by our bodies, personal, and can often be fooled into not serving our interests. The content of your conscious thought is formed and conditioned by what happens under the conscious radar in your intuitive mind. A full ninety-eight percent of your brain activity is unconscious, including how you react to the threat of climate disruption. This is true for everyone.

Cognitive Biases and Cognitive Ease

Our minds are conditioned to quickly bring consistency and coherence to situations for which we have incomplete information. We disambiguate, fill in relevant details, and make lots of fast decisions about context and relevance. Our worldview helps us make sense of things. It's what gives us the confidence to act. To come up with our worldview, we use a bunch of cognitive shortcuts (or heuristics) that bias our thinking. We are not disembodied intelligence, but we do have the cognitive biases of creatures with a background, interests, and emotions.[38]

There are three basic rules of cognitive engagement. First, our emotions play a big role in disambiguating and making sense of the world. Without an emotional trigger to guide our thinking and curtail the near-infinite ways in which we can interpret an event, we would be paralyzed or perennially confused. Second, we typically come to the scene with an opinion about *everything*, whether or not we're aware of it. According to Daniel Kahneman, "The normal state of your mind is that you have intuitive feelings and opinions about almost everything that comes your way."[39] Third, we avoid cognitive effort. We naturally prefer the fast, automatic processing of our intuitive processor over the slow, self-reflective work of conscious thought.

When we form opinions, we balance explicit information and emotional interest, but emotions dominate. Our opinions aren't always conscious; they can exist as a potential response to something. Sometimes we don't even know we have an opinion about something until we're asked! But once we've formed an opinion, we're generally very confident about it. We don't often make the effort to override our first response.

Perhaps you're watching a baseball game. You don't know much about the new pitcher, having watched only a few throws. Someone asks you about the player's chances that season. Odds are you'll have an opinion with some degree of confidence. Perhaps you notice the pitcher has thrown two strikes in a row, and he's on your home team. But you also remember he was the replacement for your favorite pitcher last year, and you notice he's wearing his socks over his uniform, a habit you unconsciously despise. Unless you make a real effort, the emotional content will drive your opinion. You won't like his chances. Self-reflection can change your answer, but we avoid cognitive effort if we can.

Now someone asks, "Do you believe in climate disruption?" This is a more complex question than the one about the pitcher since it involves science, political arguments, the IPCC, and all sorts of other stuff. The first thing our intuitive processor will do if it can't find a good answer is "find a related question that is easier and . . . answer it."[40] We simplify rather than make the effort to work it through. We *unconsciously* answer an easier question, "how do I feel about climate disruption?" The answer to that question is pretty simple: it's scary. Our emotions then make the call, sending up a don't-believe-it signal. Denial is our default position.

Anyone who has become engaged in this issue (myself included) had to go through long periods of deep reflection to get past this first emotional response and realize just how bad things are. Accepting the truth of climate disruption is not pleasant; it provides no emotional succor, nor comfort in the dead of night. Climate denial is a natural belief that must be dislodged in favor of acceptance. That takes effort. Unfortunately, there's a lineup of other cognitive biases laying in wait for anyone who wants to try.

First up is confirmation bias. When we read the paper, watch the news, or surf the web, we give greater weight to evidence that confirms what we already believe: "people . . . seek data that are likely to be compatible with the beliefs they currently hold."[41] I've found myself grasping onto articles that give the smallest bit of hope, knowing it makes the flimsiest of cases against a mountain of counterevidence. We all do it. Groups who seek to foment disbelief exploit this trait by laying out tempting cognitive treats for us. Confirmation bias ensures we're predisposed to give their anti-scientific nonsense much more credit than it deserves.

Next up is the affect heuristic, meaning our likes and dislikes dictate our opinion about something. It's the dominance of conclusions over arguments. If we like the conclusion, we believe the argument. If not, we don't. This is most pronounced when emotions are involved. If you're a Democrat and like President Obama, then his arguments about health care make sense, and if you're a Republican and don't like Obama, they do not: "Your political preference determines the arguments that you find most compelling."[42] We don't form a belief by following an argument. Rather, we judge an argument's validity based on how much we like the conclusion. Climate disruption is scary, so we're likely to see arguments in favor of its truth as invalid. We might also dislike the political ramifications of climate disruption, for example the need to constrain the market, which leads us to the same conclusion. This explains why belief in climate disruption is so split across party lines.

Would you believe a good-looking weathercaster or a pointy-headed climate scientist with a large ego and impatient manner? Most of us are unconsciously biased toward the weathercaster, even though he has no real credibility on climate science. This is an instance of the halo effect: "The tendency to like (or dislike) everything about a person — including things you have not observed."[43] We tend to trust people with social grace because we like them. That explains the winning smiles on all those victorious politicians! Fox News knows this, and so does the fossil fuel industry. Pretty talking heads are effective, even up against Nobel laureates who study climate science. Talking heads are chosen for their ability to trigger the halo effect. Scientists are chosen because of a career filled with equations, peer-reviewed papers, and deep thinking. The halo effect makes it no contest: when we seek the truth on climate science, we are distracted by personal charm.

Then there's the peer group effect, meaning we put more weight on the views of our peers than anyone else, including experts. When oil executives or business leaders surround themselves with like-minded people, they become trapped in an echo chamber that rein-forces their beliefs. The same thing happens when we form singular communities on the Internet. The peer group effect makes it more important than ever for politicians to allow dissenting voices into their inner circle on a regular basis. Canadian prime minister Stephen Harper is notorious for hearing the views of only a few like-minded people. So was George W. Bush. The peer group effect explains why ideologues become so immune to what seems, to an outsider, obvious empirical counterevidence.

In an ironic twist, we are also prone to loss aversion, meaning we're more likely to avoid a loss than to strive for a gain. There are lots of experiments with betting that show this bias in some detail: the psy-chological loss of a given amount of money is greater than a psycholog-ical gain of the same amount. In other words, "losses loom larger than gains."[44] Since climate disruption brings huge losses, you'd be forgiven for thinking loss aversion bias is *toward* acceptance of its truth—but this is not the case!

That's because we are also short-term in our thinking. A short-term loss trumps a long-term threat. Responding to climate disrup-tion requires short-term losses for many of our existing institutions. This explains our inherent conservatism and a defense of the status quo even in the face of long-term catastrophic loss. According to Kahneman, "Loss aversion is a powerful conservative force that favors minimal changes from the status quo in the lives of both institutions and individuals. . . . [I]t is the gravitational force that holds our life together near the reference point."[45] The irony is loss aversión biases our thinking and action toward the long-term catastrophic losses associated with climate disruption.

Our minds play all kinds of games. These cognitive shortcuts let us avoid careful, effortful thinking. The bottom line is we seek what is called cognitive ease. We don't want to expend much cognitive effort but do want confidence, quick answers, and a good mood: "Good mood and cognitive ease are the human equivalents of assessments of safety and familiarity."[46] Our minds will dismiss evidence, ignore information, and pull all sorts of stunts to find cognitive ease. Author Margaret Heffernan explains, "This is where our willful blindness originates: in the innate human desire for familiarity, for likeness, that is fundamental to the ways our minds work."[47]

Belief in climate disruption provides us with the opposite of ease. It causes discomfort, worry, and distress. It's only natural we find ways to avoid it. Our great, collective intellectual challenge is to face that belief—not just believing climate disruption is true, but the more difficult task of facing up to how advanced and dangerous it has become.

Thinking About Climate Disruption

George Orwell's dystopian novel *1984* coined the term doublethink, described as "the power of holding two contradictory beliefs in one's mind simultaneously, and accepting both of them."[48] For Orwell, it was possible to permanently maintain that condition. Indeed, in *1984* that skill was a precondition of proper social integration, but it takes more and more effort for our conscious minds to doubt the truth as evidence becomes overwhelming. At some point, drilling for oil in a melting Arctic becomes less ironic and more alarming. Sustaining doubt becomes hard work. As evidence mounts and more credible institutions ring the alarm, it becomes impossible for our cognitive biases — strong as they are — to continue to hold the gates against a growing flood of information. Eventually something must give.

In the real world, cognitive dissonance describes the discomfort we feel when faced with two such incompatible beliefs. We'll try to avoid that discomfort as best we can and will seek cognitive ease through cognitive bias. But as the world warms and the weather gets nastier and more extreme, our experience of cognitive dissonance grows to the point of discomfort. Cognitive dissonance is a necessary step to get past climate denial because climate disruption contradicts a whole suite of beliefs that are fundamental to our shared worldview.

As we go through life, we typically surround ourselves with stuff we like: people who agree with us as well as sets of ideas that make us comfortable and play to our strengths. We become socially and professionally committed to a particular view of the world. Out of that set of experiences, we build our intuitive processor — a realization of our worldview, through which we filter new experiences, evaluate threats, make judgments, and form new beliefs. Much of that worldview is in stark opposition to accepting climate disruption for the threat it is: climate disruption is, quite literally, hard to think about.

First off, all of us believe the future will be something like the past. That's the basis of any cognition or action. It's also customary to believe the lives of our children will be better than our own and that the future brings progress. For much of history that's been the case. A sense of progress is fundamental to Western economic thinking, and it's a central pillar of the American dream. The upcoming climate storm throws into doubt a continued rise in our collective standard of living, even the notion of a relatively stable global economy.

Related to this is the idea that the economy can expand indefinitely. It forms the basis of modern economic thinking. When the global think tank Club of Rome proposed in the 1970s that there were limits to economic expansion, they were essentially laughed out of the room. At the very least, climate disruption puts an end to the notion that we can continue to power our growing economy with fossil fuels. More poignantly, the size of the storm that brings droughts, rising oceans, food insecurity and wars over water calls into question the ability of our economy to continue to grow at all. Continued progress and unlimited economic growth are very difficult ideas to give up.

Many of us also believe humans can solve whatever problems emerge, including climate disruption. We've done well so far, so why should the carbon problem be any different? Treating climate disruption as a unique threat insults our very nature and diminishes the magnitude of our collective journey from the African savanna to industrial civilization. This degree of self-belief is an argument by induction: we've risen to challenges many times in the past, so the odds are good we'll rise to this one too.

But the truth is we've faced very few existential global threats in the past. A depleting ozone layer and a potential nuclear winter are the only two I can think of. The sample size for showing we solve all problems is tiny, and there are few conclusions we can draw from them. Furthermore, negotiating a treaty on air-conditioners or nuclear triggers was a piece of cake compared to carbon emissions. Climate disruption is unique in scope, scale, and complexity—it goes to the heart of the global economy. The solution involves nearly everyone. The truth is we've never faced a threat this complex before. The belief that our ingenuity will spring into action provides a wonderful sense of hope. It's hard to give up, but hope is all it is.

For some groups, like those on the political right of the spectrum in North America, cognitive dissonance is more acute because the implications affect a much larger set of pre-existing beliefs. The state is the only institution that can set the market signals we need to solve the climate problem. The state as the solution is anathema to many with right-wing views. Worse, once you acknowledge the necessity to price carbon emissions, then you inevitably have to deal with related (and difficult) moral issues. How will the poor be affected by higher energy costs? How do you accommodate out-of-work coal miners? How does the developed world acknowledge its much higher per-capita emissions?

All these issues fly in the face of the neoconservative view of free markets, small government, and U.S.-style fend-for-yourself independence. Yale's Cultural Cognition Project has found that politics and worldview explain "individuals' beliefs about global warming more powerfully than any other individual characteristic."[49] That's because belief in climate change runs counter to core economic beliefs, which act as ideologies that create cognitive dissonance.

I think one of the most fundamental traits we'll have to give up is our hubris. We generally (if implicitly) see ourselves as infinitely clever, as masters of our domain and in control of our destiny. We perceive nature as something we bend to our will, a view that traces its modern roots to Sir Francis Bacon. Technology is how we exert our strength and intelligence; it's an expression of Nietzsche's will to power. We've good reason to think highly of ourselves — we put a man on the moon and split the atom! But climate disruption requires a more humble view. We have enough strength to upset the apple cart, which confirms our hubris, but we might not have enough to set it right again. Nature is in control, not us. It is to nature we must submit our economy, our infrastructure, and our civilization. That requires a much more humble view of ourselves.

Reactions to cognitive dissonance can be rationalization, anger, or outright dismissal. Many on the far right — like Glenn Beck and

Newt Gingrich—dismiss climate disruption as a socialist plot. Their rationale is because the solution involves the state and those who advocate state-led economic action are socialists, it follows that the whole thing's a play for power by a bunch of watermelons (green on the outside, red on the inside).

But as temperatures rise, the scam has to become more and more complex, until it just sounds ridiculous. The rationalization engine goes into overdrive. It's a conversation with Brian all over again. The right wing's worldview is deeply threatened by climate disruption. To break out of that cycle we need a strong, respected leader from the right—a modern-day Thatcher—to take the lead. Perhaps the fears of the right wing are alleviated when one of their own champions a market-based solution.

People express a lot of anger at climate scientists. Threatening emails are daily occurrences for those who speak out. You only have to read the comments section underneath any online news article about climate disruption to get a sense of the high emotions climate science provokes. The truth about the climate can steal cognitive ease and replace our sense of familiarity with one of fear. It threatens our families and way of life. It's a belief we do not want to let into our cognitive environment. It's natural to feel anger at this threat, however misplaced or unhelpful that anger may be.

These reactions—dismissal, rationalization, anger—are a good beginning, but we need to start a new kind of conversation, one that gets past our biases, fears, and ideologies.

WAKING THE FROG: STARTING A NEW CONVERSATION

Instead of fighting our shared cognitive biases, a smart communications strategy seeks to take *advantage* of them. Imagine a charming, telegenic champion of an era of renewed prosperity that is underwritten by a job-creating economic stimulus and powered by abundant low-carbon energy sources. Instead of fear or confusion, our first reaction to climate disruption is one of possibility—a brave new world of clean energy abundance and sustainable economic growth. Our default becomes acceptance. The charming spokesperson's vision is boosted by the halo effect, and because we like the conclusion—climate action leads to economic stimulus—the affect heuristic works with us. When civic leaders of all stripes sing this tune, confirmation bias reinforces acceptance. And when that conversation spreads into our daily lives, peer group bias defends our new belief. An energy moon shot (see page 169) to an abundant, clean future speaks *to* our biases, not *against* them. Standard fare in professional marketing and PR campaigns.

It's also possible to imagine a strategy that gets around those biases. In our culture, artists are the people best able to articulate truths in ways that skirt rational thinking. Storytellers, C.S. Lewis said, carry meaning in a way that rational truth-tellers cannot. "For me," the novelist wrote, "reason is the natural organ of truth; but imagination is the organ of meaning. Imagination, producing new metaphors or revivifying old, is not the cause of truth, but its condition."[50]

Cape Farewell taps into that renewable resource. Founded in the United Kingdom by photographer, artist, and all-around buccaneer David Buckland in 2003, Cape Farewell brings artists and scientists together on journeys to the front lines of the changing climate—often by boat to the Arctic. Over a number of weeks, those very different types of thinkers share stories, data points, and songs. Upon their return, new ways of thinking and communicating about climate disruption are unleashed into our culture. As Buckland is fond of saying, "Climate is culture, mate!" Cape Farewell's Canadian opener was the art exhibit Carbon-14 at the Royal Ontario Museum in the fall of 2013. My own contribution (a collaboration with Buckland) was a bullet made of $180 worth of silver, representing a price on carbon that I see as the "silver bullet" solution.

Cape Farewell's mission doesn't lack ambition. They want nothing less than to change the zeitgeist and embed a response to climate disruption into the very genes of our culture. The idea is simple: scientists, brilliant as they may be in understanding the inner workings of the world, are terrible communicators. It's the artists who tell the stories that linger, sing the songs that can lift our hearts, and make the pieces of culture that last through the centuries, so put artists and scientists together in an environment that makes them think, share, and interact. Let the scientists inform the artists, and have the artists inform the world. Cape Farewell's global art projects — most recently Carbon-12 in Paris, Carbon-13 in Texas and Carbon-14 in Toronto — are the tip of the iceberg.

What really engages the broader public is the way participating artists — like singer-songwriter Feist and novelists Yann Martel and Ian McEwan — incorporate what they've learned into popular culture. "The pressure of our numbers, the abundance of our inventions, the blind forces of our desires and needs are generating a heat — the hot breath of our civilization. How can we begin to restrain ourselves?" wrote novelist McEwan after visiting the melting Arctic ice on a Cape Farewell voyage. "We resemble a successful lichen, a ravaging bloom of algae, a mold enveloping a fruit. We are fouling our nest, and we know we must act decisively, against our immediate inclinations. But can we agree among ourselves? We are a clever but quarrelsome species — in our public debates we can sound like a rookery in full throat. We are superstitious, hierarchical and self-interested, just when the moment requires us to be rational, even-handed and altruistic."[51]

Cape Farewell is a means to find a new conversation, a way of getting past cognitive biases. Information by itself is not enough. Telling people how bad things are getting just makes them more defensive and builds feelings of helplessness. We all have an emotional stake in disbelieving the worst, and as clinical psychologist and psychoanalyst Todd Essig wrote in *Forbes*, "You can't smash emotions with reason."[52] The paradox is that unless we understand how dire the situation is, we won't react appropriately. But understanding how dire it is causes paralysis. The way out can only be through a new conversation. Artists quite naturally shoot for the heart and get past our biases, defences, and feelings of fear.

There are lots of other ways to start that new conversation. All civic leaders owe it to themselves to play a part. We can't stand back and allow this conversation to be driven by the talking heads on Fox News, thinking that somewhere, somehow things will turn out all right. Our religious leaders can step up to the plate. Some are, but they can do more. A conversation around faith, and obligations to a larger purpose, might sneak past our mental defenses.

But where our defenses are most likely to come down is at home. When your kids ask at the family dinner, "What are we going to do when it gets hot?" or "What did you do when it started getting hot, mom?" don't avoid the question. Use it as a starting point to engage them. Answering to the simple moral positions that only the young can really articulate is not easy, but if we're listening, that moment can provide exactly the motivation we need to disengage our defenses and face the truth.

Suggested Reading

Susan Blackmore, *The Meme Machine* (1999)

Andy Clark, *Associative Engines: Connectionism, Concepts and Representational Change* (1993)

Hubert Dreyfus, *What Computers Still Can't Do: A Critique of Artificial Reason* (1992)

Daniel Kahneman, *Thinking, Fast and Slow* (2011)

George Lakoff, *The Political Mind: A Cognitive Scientist's Guide to Your Brain and Its Politics* (2009)

CHAPTER 3
Complexity and the Myth of the Free Market

"How many Chicago School economists does it take to change a light bulb? None. If it needed fixing the market would have already fixed it."
— Joke among non-Chicago School economists

Left to its own devices, the market won't solve our climate problem. That much is clear to anyone who reads the daily business section with an eye on atmospheric carbon counts. There is no letup in fossil fuel infrastructure spending; at the time of writing, carbon counts are 397 parts per million and rising another 2-3 ppm per year. We must decisively and radically interfere in the market if we are to move away from what are still plentiful and reliable fossil fuels. The market's powerful and creative force can be harnessed to build a new low-carbon economy. Free of constraints it will cause every economically viable chunk of coal in the ground to be burned. That's suicide.

A modern myth hinders our progress on marshaling market forces against climate disruption: the sanctity of the free market. In its starkest form, the myth informs the neoconservative's belief that, when it comes to the market, "the freer, the better." I call it a myth because there is no such thing as a free market. There never was, and there never can be. We are distracted from action on the climate by a logical fallacy, a paradox.

The myth takes a number of more subtle forms: only with government out of the way do we maximize growth; heroic entrepreneurs, free of constraint, drive all wealth and job creation; an efficient economy requires that people and corporations be unconstrained in pursuing their self-interest; regulations are like friction, hindering growth and job creation; and the unparalleled wealth of modern society is entirely due to free markets. What ties these myths together is the idea that "freedom" logically and pragmatically precedes "market": the market is judged as effective, based on how much freedom it allows.

There is *some* truth to each statement. The market is a powerful force, unmatched in creative potential. Entrepreneurs are key to innovation and wealth creation. Self-interest is a powerful motivator. Bad regulations can be stifling. Relatively free markets are a prerequisite to efficiently creating and maintaining wealth. It is one thing to embrace the market, and another to say it must be left alone. That's like saying fire is useful, let it rage!

One might be forgiven for thinking that the recent financial meltdown—caused in large part by a deregulated financial sector[53] that suddenly found itself free to act with unrestrained fervor and that was only barely contained by massive government action—would put that myth to rest. But with varying degrees of commitment, the political and business consensus in North America remains that markets should be left alone if they are to be effective.

Those who champion the free market often speak of it as a naturally occurring element like a tree or a tiger—a so-called "natural kind," in philosophical jargon. Market forces are treated with reverence, akin to natural forces like the flow of a river or the force of the wind. Our task is to understand its natural laws and ensure our own are subservient to it. Traditional economics provides cover for this view by importing the language of the physical sciences, likening the market to a physical system that seeks equilibrium. Market equilibrium is a mythical place that makes everyone happy and can only be reached if we leave the market alone.

But the market is not a natural kind, and it doesn't operate at anything like equilibrium. Part artifact and part natural, it's a hybrid. We invented the rules of the game: currency trading, the banking system, initial public offerings, contract law, stock markets, credit default swaps, insurance, and even the idea of trading one thing for another. It's also part of nature: us, our psychology and creativity, the resources we use, and the ecosystem we depend on. The global economy is a highly complex system that operates far from equilibrium, constantly evolving and reinventing itself. The modern language of dynamic systems is more useful than the old math of equilibrium. This more nuanced view has implications for the best way to shape the market to solve climate disruption.

Let's get past the myth of free and emphasize market. Markets are the most powerful and creative social tools in the history of human civilization, but they've never been free. Getting off carbon takes more than mere tinkering; we need a wholesale reconstruction of global energy flows in a very short time. We waste crucial time staring down such a purist view of the market. Our task is to engineer new rules to maximize the market's creative potential to solve our climate problem.

How Free Is Free?

With apologies to my libertarian friends, there is no such thing as a free market. It's a logical impossibility. If you want to see what a truly free market might be like, look no further than the hilly tribal areas between Pakistan and Afghanistan or to Somalia, where there hasn't been an effective government for a couple of decades. No taxes, few regulations, no government-imposed banking regulations, and so on. These are a neocon libertarian's dream landscape, ideal places for the mythical free market to emerge. It hasn't.

Getting rid of all the rules doesn't get rid of structure, at least not for long. It certainly doesn't result in the wealth-generating, productive, and entrepreneurially dynamic economy we want. The lack of rules creates a vacuum that is inevitably and quickly filled with structure imposed by nefarious groups able to play strongman. And so no rules results in the strongman's rules and the kind of highly constrictive society few of us would choose to live in.

In a more developed marketplace, the strongman is the monopoly, oligopoly, or too-big-to-fail gambling joint (occasionally recognized as a Wall Street bank). Unrestricted markets allow the equivalent of the schoolyard bully to dictate the rules of the playground and rob you of your lunch money. At the turn of last century, the railroads replaced farmers as strongman, then it was Standard Oil followed by the financial gluttony and robber barons of the roaring twenties. The excesses of Wall Street culminating in massive direct bank bailouts followed by the more nuanced heist of the Federal Reserve's trillions[54] is only the latest round in a game that's been played out countless times before.

Humans do not, and cannot, live without structure. In the absence of structure chosen by a democratic government, other kinds of hierarchies and systems of power arise, be they religious, corporate, criminal, or otherwise. Lack of formal structure does not result in a theoretical maximization of freedom. It just means other ways of organizing around power. So let's not pretend there is some ideological purity about freedom. It is a relative term.

A functioning marketplace requires at a minimum the rule of law, a system of courts, regulated banking, currency, trade regulations, anti-monopoly regulations, and so on. Closer to home, we add other market constraints that reflect our shared moral values. Scary-looking guys with beards, motorbikes, and leather jackets are not allowed to sell white powder to schoolkids. Why not? Because we've decided it's bad. Instead of talking about the free market, we might want to talk about an effective or productive market.

I am not saying freedom in the marketplace is bad, only that it must—by force of logic—be applied with a matter of degree. Market forces are powered by the minimum structure required for a stable marketplace to exist. Beyond that, how we best harness those forces to solve a collective problem or achieve some common goal is an interesting and nuanced debate. It's a matter of balance: freedom for market forces to act, and structure to support or guide those forces. We've been working out that balance since the birth of market economies.

Look no further than the Old Testament for an agricultural market-stabilizing strategy: the Bible tells us to build a central grain reserve to balance fat years with lean ones. The massive yields of fat years cause food prices to plummet. Because the farmers who needed support during lean years may go bankrupt, any lean years that follow can be doubly disastrous.

In a practical sense, governments have long been the bankers of last resort. The recent financial collapse saw banks of all shapes and sizes lining up at the Federal Reserve's window, cap in hand, to take advantage of the massive liquidity injections that were required to stabilize the U.S. banking system. Left to its own devices, Wall Street's unfettered greed and self-interested pursuit of profits would have brought down the entire global financial system. That's where an unrestricted market takes you, even in a society where the basic building blocks have been in place for generations.

Some argue that the financial collapse was caused by government action, as it sought to enable lower-income Americans to own their own homes through Federal guarantee programs like Fannie May and Freddie Mac. There is a grain of truth to it—surely, a rush to sub-prime mortgages exacerbated an existing problem—but the argument reeks of self-serving rationalization. There were far too many disingenuous players, both at the retail level, where bad deals were being served up for huge commissions, and at the investment bank level, where the re-bundling of mortgages was clearly dishonest.[55] If you want to see how business acts when unconstrained by regulatory oversight, look no further than the unsavory salespeople and big boardrooms behind the sub-prime mortgage fiasco (or Enron or Bre-X or Long-Term Capital Management).

Starting in the 1980s, the Washington Consensus—a basket of free-marketeering policy prescriptions for developing countries—dominated such institutions as the World Bank and International Monetary Fund (IMF). Chicago School neocons, acolytes of economist and free-market guru Milton Friedman, played God with entire economies in the developing world. Experiments in free-market extremism were played out country by country, from Poland to Argentina. As those countries required recapitalization of their large (and often imposed) sovereign debts,[56] the institutions sent in to save them attached severe free-market reforms to their aid. State enterprises were sold off, public expenditures were slashed, regulations were tossed, and capital and commodity markets were opened up to foreign competition. More often than not, they led to economic disasters: currency and capital market instability, mass unemployment, and the misery of disappearing social safety nets.[57]

The neocons' experiments delivered bad news: empirical evidence indicated fully unrestricted markets are destructive and unstable. Some responded that the market reforms were not free enough, so the experiment was flawed. Again, this doesn't smell right. A theory is empty if counterevidence can always be ruled out. Free market theory is meaningless if, upon a failed experiment, it's ruled the market was not free *enough*. That's not a theory; it's blind ideology.

Even hard-core Republicans, like former governor of Florida Jeb Bush, capitulate on free-market ideology when it comes to practical matters. Insurance companies have long recognized that climate disruption is bad news. They're the ones who pay out when hurricanes devastate the coastal towns in Florida and help Australians recover from floods and fires. They've got the data and the financial incentive to take a warming world seriously. The result? They stopped offering insurance to homes in Florida. That meant homes couldn't get re-mortgaged or sold. The Floridian economy relies heavily on the housing market, so Jeb Bush offered up the state government, through the Citizens Property Insurance agency, as an insurer of last resort.

What happens when the next hurricane devastates Floridian real estate? It could bankrupt the Florida state government, since the claims could be as high as the entire annual Florida state budget.[58] That means the federal government will step in to bail out the state government, which is only there because the private sector would not take on the risk. There is no free-market solution to these sorts of disasters, only the state can provide a backstop. That's what underwrites the modern nuclear industry as well as what provides protection from many terrorist attacks.[59] Government will certainly underpin future climate disasters.

The irony is that Jeb Bush is a climate sceptic, pushing hard to keep the government out of the fossil fuel energy industry in the name of free enterprise. So even while he hinders action on climate disruption in the name of free markets, he can only protect his economy from its ravages by relying on two levels of government to intervene in the market by pledging their balance sheets.

It's reasonable to wash away the fundamentally incoherent ideology of free markets. Once we acknowledge that structure is required, it's a matter of asking how much structure is appropriate, and the answer to that question depends on what you're trying to achieve. If you want a Russian-style oligopoly based on resource extraction, you get one answer. If you want a knowledge-based and innovation-driven economy like Silicon Valley, you get another. And, of course, if you want a sustainable global economy that's decoupled from carbon emissions in time to avoid catastrophic climate damage, you get yet another answer.

State-Enabled Wealth Creation

Much of our modern economy—from the aerospace industry to the Internet and from the automotive sector to the steel industry—exists because previous governments strategically intervened in the market.[60] Wealth creation is maximized by a balance of state and private activity. State interference can help or hinder, depending on the details, but the notion that state action is always bad is sheer nonsense. Today's captains of industry owe a great deal to activist governments, whether they acknowledge it or not.

As we've seen, at the very least states create the stable market conditions that precede any healthy economy: laws, currency, courts, etc. While theorists might argue about the benefits of further stabilizing actions—like traditional Keynesian stimulus-spending during a recession[61]—as a practical matter, governments have had to take action to save captains of industry from themselves. The recent bank bailouts were only the latest in a long series (having been preceded by the Savings & Loan scandal of the 1980s and the Great Depression for starters). The North American auto industry was recently brought back from the edge of bankruptcy after CEOs arrived in Washington on bended knee (and private jet). That governments must provide stability has been proven time and again in crisis, but here I want to make a different point.

Historically, the state has also been active in enabling wealth creation. As U.S. economist Joseph Stiglitz puts it,[62] the American markets "were not left to develop willy-nilly on their own; government played a vital role in shaping the evolution of the economy. . . . The federal government played a central role . . . in promoting American growth."[63] The state has long been a partner in seeding new industries and helping create the conditions in which they prosper. A partial list of wealth creation includes basic research to unlock new technologies, building enabling infrastructure, putting up protective trade barriers for nascent industries, direct subsidies, and providing initial demand to lower cost and generate scale.

For example, in 1956 the U.S. federal government implemented the Federal-Aid Highway Act. Thus was born the massive U.S. interstate highway system, and the era of the great American road trip began. In one fell swoop, the U.S. automotive sector was given a massive shot of adrenaline. While the act had a number of effects—giving the U.S. military the ability to move troops effectively across the country, increasing trade and tourism—there's no doubt it was a boon to the U.S. auto sector. That sector would almost certainly not have become as large as it did without those highways. The federal government created the infrastructure that unlocked a half-century of economic dominance by Detroit.[64] That wealth was mirrored in Canada's auto sector, which was backed by our own highways and a strong push by Lester Pearson for the Auto Pact.

The aerospace industry has always relied on military demand. The nuclear industry, though now in the doldrums given public pushback on perceived safety risks, was entirely a product of publicly funded wartime development. Even the Internet—which generates enormous wealth across dozens of countries and is a paradigmatic example of entrepreneurs jousting among each other in the meritocracy of the free marketplace of ideas — is the result of a public and private collaboration. The computer industry was not the result of lone entrepreneurs working away in their garage. Hewlett-Packard, a bastion of Silicon Valley, could not have existed without the long history of publicly funded research into computing that preceded it and the strong initial demand for the microchip from the military. Silicon Valley did not emerge spontaneously from the orange groves of California but began as collaboration between Stanford University and the private sector.

The myth of the free market is often personalized in an image of a heroically independent, wealth-creating entrepreneur. From the original Ford churning out cars for the masses to the giant-killing Steve Jobs and his single-minded pursuit of design perfection that became Apple, entrepreneurs who beat the odds and create enormous wealth are often (if implicitly) taken as the standard-bearers of market freedom.

I don't deny the wealth-producing effects of the entrepreneur's work — I am one — nor the effort it takes to be successful, but entrepreneurs, from auto magnates to Internet billionaires, are able to achieve what they do only because of a surrounding structure of civil society and a long historical arc of technological development that itself relied on the public sphere. Relying on an educated public from which to hire employees or direct public investment in research and infrastructure or even direct support for emerging industries, entrepreneurs do not act alone. The structure they leverage did not arise from nothing in the vacuum of a free market, but largely out of actions taken by the state. Bill Gates could not have built Microsoft, nor Steve Jobs Apple nor Mike Lazaridis Research In Motion, without the publicly supported education and research that preceded their arrival. The mythical independence of the self-made entrepreneur is a powerful story that increases our unwillingness to interfere in the market.

Admittedly, the private sector is uniquely able to bring these technologies into commercial production and expose them to the rigors of the marketplace—and the market picks the winners, not the public sector. My point is softer: there has always been a role for the public sector in wealth generation. The details can be complex, and there are assuredly as many white elephants as there are successes, but the wealth we take for granted today, across lots of sectors, was not conjured up exclusively by the magic of the private sector.

Perhaps we'd have generated all that wealth without government involvement. Maybe we'd be even wealthier if there was no government interference. I doubt it. When evidence of the effectiveness of public-private collaboration rests in some of the largest, most profitable, and longest-lasting industries, the burden of proof is not mine to bear. That sort of counterfactual argument—what might have been had the world been different—is disingenuous because it can disavow *any* evidence. It's like the arguments about the failures of free-market reforms we saw above: a theory that can't be disproven because the world could have been otherwise is unhelpful, at best.

The irony is that the fossil fuel industry itself, which often fights tooth and nail against market interference that supports clean energy, wouldn't exist without massive, long-term, and ongoing support from governments around the world.[65] Historically, the oil industry required governments to underwrite and protect its investments. The industry is still the recipient of hundreds of billions of dollars of direct subsidies, along with indirect subsidies such as standing militaries to protect capital investments and trade routes, and we haven't yet come to the massive indirect subsidy of free carbon emissions.

From Equilibrium to Dynamic Systems

A lot of ink has been spilled developing the theory that lurks behind the modern myth of the free market. Starting with Adam Smith's invisible hand and culminating in the Chicago School's triumph of free-market fundamentalism in the late twentieth century, philosophers and economists have built a huge intellectual edifice spanning hundreds of years. But it's outlived its usefulness. The view of the market as a kind of natural force best left alone sits on out-of-date mathematical ideas about equilibrium that were inherited from ancient physics.

It's time to modernize our view of the economy. Today's understanding of complex, dynamic systems far from equilibrium—like weather patterns, stock markets, and biological systems—is much closer to real economic activity. It has lessons for how we might harness market forces to solve the carbon problem. Markets can be too free, but interference can be too blunt. A more nuanced understanding of market complexities should guide smart policy.

The Traditional View: Free to Find Equilibrium

In a nutshell, the traditional view of markets contends that markets must be free to find their natural equilibrium and that equilibrium is where supply meets demand, prices and allocation of resources are optimal, and everyone's happiness ("utility" or "welfare") is maximized.[66] Equilibrium is utopia! And self-interest gets us there, acting as a force that moves the system to equilibrium, just like gravity makes water flow downhill. Prices communicate this force throughout the market, so a free market is one where everyone pursues their self-interest and the "invisible hand" magically sets prices and distributes resources to everyone's benefit.

If the market isn't free, it can't find its equilibrium. That means everything from prices to happiness to wealth is suboptimal. Hence, the freer, the better — for everyone. Want to maximize wealth? Let the billionaires play. Want to lift people out of poverty? Open up markets. Want to efficiently distribute goods and services? Cut regulations to zero. Want to protect the environment? Leave the markets alone. The unhindered pursuit of self-interest, the defining trait of a free market, is raised to the status of moral imperative. Seem a bit self-serving? It is. But as economics got more sophisticated, such simple criticisms were easily brushed aside by an onslaught of academic language.

The story goes like this: Adam Smith's insight was to see that the economy gets more productive as people specialize. As such, people have to be able to trade with each other. Since people are the best judge of their own happiness, fair trades are the only ones people will make. If a trade isn't fair, the argument goes, a person wouldn't make it (a circular argument that is invalid).[67] Thus, people must be free to pursue their own self-interest. The profit motive (and competition) drives efficiency and maximizes production and happiness. Freely competitive markets get morally grounded. "It is not from the benevolence of the butcher, brewer, or the baker that we expect our dinner, but from the regard to their own interest."[68] This is the famous invisible hand of competitive markets.

A French economist and statesman, Jacques Turgot, pointed out that producers eventually stop putting more money into production since they get a bit less additional output for each additional input. Think apples: the first additional worker gets you a lot more apples, the second a bit less, and so on. That's because there's limited land, you have to organize the workers, etc. That's the law of diminishing returns on production. An English philosopher, Jeremy Bentham, did the same for demand. For every additional apple we eat, we get a little bit less happiness (we start to get sick of apples) and are willing to pay less. That's the law of diminishing returns on utility. Put these laws together and, voilà, the price of stuff brings about a balance between the two.

This sounds sensible, and there is a great deal of truth to it. Economics as a kind of philosophy provided good rough generalizations, but then it got math and began to look and act more like a hard science. This made it seem more credible, as if it were describing a real thing—like a tree or a tiger. Equilibrium and self-interest gained the status of physical things. It seemed best to bend to these forces of nature.

During the 1800s, the physical sciences went through enormous advancement. Giants like Faraday and Maxwell arrived on the scene, and a new math called differential calculus was invented. This new math accurately described all sorts of things that were previously quite mysterious. A grand age of science had begun, and everything from the motion of planets to electricity to the waves on the sea were captured and explained in great detail. Economics got in the game. The "big idea" was economic activity would become as predictable as balls rolling down a ramp.

Lots of people jumped in on the action. French mathematical economist Leon Walras borrowed equations from physics to specify the notion of equilibrium in how people traded stuff. Like a ball rolling to the bottom of a bowl, we'd trade until prices hit equilibrium. To make the math work, there could be only a single point of equilibrium. English economist and logician William Jevons took literally the idea that self-interest was a force, like gravity. He surmised that it acted on differences in people's preferences—a potential energy—through trading. He formalized the idea of constrained optimization, where the equilibrium we reach optimizes our own happiness within the constraints of limited resources. Italian economist Vilfredo Pareto put the finishing touches on the picture by arguing that the equilibrium this system would find happens to be the best possible state of affairs. That's because people would keep trading only until they could gain no more happiness. That's Pareto optimality—the promise that the free market lands on a fair and equitable spot.

A few more bits got added. People's preferences could be seen by the choices they made in the market. Values and happiness, until then ineffable, were subject to empirical measurement. Optimality and equilibrium got matched over multiple markets—as one item ran short and got expensive (like labor or fossil fuel), others could be substituted (like automation or wind power). Price acted as the central nervous system. Growth was accounted for by heroic entrepreneurs, whose innovations were thought of as adding the equivalent of energy to the system. Productivity meant equilibrium was dynamic. Because technology is based on knowledge and knowledge is unbounded, economic growth is also unbounded.

Add it all up and out pops the central lesson: undistorted competitive markets, with people left to pursue their own self-interest, will automatically arrive at an ever-growing dynamic equilibrium guaranteed to maximize everyone's happiness. The closer we get to perfectly free markets, the closer we are to utopia.

Not only was this very good news—particularly if you happened to be a billionaire who wants unadulterated pursuit of personal wealth—but a ton of complex math backed up the story, the same math that proved so good at predicting planetary motions, pumping pistons, and, yes, balls rolling down inclines. So the story must be true! The Chicago School gobbled it all up and applied the lessons to macroeconomic theory and public policy.

It's hard to overstate the influence economics gained as it took on the status of an empirical, mathematically grounded discipline. Anyone who wanted regulations out of the way, the state to sell off assets, or just liked and benefited from rampant capitalism now had a robust and complicated-sounding theoretical justification for their preference. Common-sense objections to the unraveling of state participation in the market were no match for published academics speaking the language of Pareto-optimizing equilibriums. Even in the face of glaring mistakes and inconsistencies—the disasters created by the World Bank as they imposed the Washington Consensus on country after country, the clear danger of increasing carbon counts, the highly volatile financial markets—free-market ideologues had the cover of smart academics and highly technical language as they took the central lesson to be "leave the market alone."

The problem with all this wonderful theory is that it only vaguely describes the real world. It's at best a rough rule of thumb, yet it pretends to have great precision and insight. The whole intellectual edifice is suspect, and the sorts of lessons one can draw are limited. There are three deep criticisms, none of which has been definitively refuted: the assumptions required to make it work are demonstrably false, it has little predictive power (a key indicator of scientific relevance), and it consistently ignores counterevidence.

Bad assumptions are rampant. For example, Walras fixed the system so there was only one point of equilibrium. He did so not because it was true, but because his dream was to make economic systems predictable. The math had to be simple. As we'll see, a signature of real-world dynamic systems is that they are complex enough to defy prediction and have multiple (and changing) points of equilibrium. A ball rolling to the bottom of a bowl is a simple picture, but it doesn't correspond to much activity of interest in the real world. It's like those physics problems students answer in Grade 10: given some massively simplifying set of assumptions, we can predict the movement of things sliding down ramps or dropping off cliffs. Throw in a bit of friction or have the thing bounce a few times, and the problem becomes intractable.

When Jevons introduced constrained optimization, he had to assume consumers were super-computers, capable of comparing every transaction they make to every other possible transaction in real time! When you buy a pound of apples, in order for it to be an efficient transaction you would need to compare the price and your enjoyment of the apples to the cost and enjoyment of bananas shipped in tomorrow from Mexico, the price and quality of a television down payment, and even the future cost of car insurance. We clearly do not, and cannot, do that. Modern models try to be more realistic in having people "satisfice" their demands—which means a rough-and-ready judgment, but it's impossible to model and really just a fudge.

A lot of other faulty assumptions are baked into the math of market equilibrium: information is perfect and flows freely and instantaneously; branding makes no difference and everything's a commodity; there are no transaction costs; people are not manipulated; monopolies don't exist; anything for which there's a shortage can always be replaced with something else; time doesn't matter; there's no institutional or psychological momentum in the system, whereby people continue to prefer the status quo simply because they prefer the status quo or don't want to stick their neck out and take reputational risk—and on it goes. There are intellectual Band-Aids for each of these criticisms. Economics twists and turns in all sorts of ways to try to accommodate real-world anomalies, but that's precisely the point: the real world is more often than not anomalous to the core theory, which tells you the core theory isn't talking about the only world that matters, the one we live in.

Maybe these are just good simplifying assumptions. After all, science and engineering do it all the time. Think of all those diagrams of swinging pendulums and rolling balls without air or surface friction and planets whose center of gravity is modeled as a single point in space. Surely we can give economics the same latitude. But simplifications in science don't work that way. The purpose of a scientific theory is to *explain* how things work, to look under the hood and examine what sorts of things cause what other sorts of things. Simplifications, like frictionless surfaces, isolate underlying physical principles. They are not imposed merely for mathematical convenience.

The lessons you can draw from an economic model are extremely limited. Those who look to Smith's invisible hand or Pareto optimization to draw the strong conclusion that markets should be as free as possible are overstepping the bounds of such a limited model. It's as if physics studied only frictionless surfaces and decreed engines need no lubricating oil.

U.S. economist Milton Friedman and the Chicago School were deeply committed to economic orthodoxy, and they took on this philosophical assault. Forget looking under the hood for causes, as long as the models reflected the relevant patterns, they were good enough. As long as it was as if people were perfectly rational, for example, what more do you want? Economics would lose its explanatory power but at the gain of predictive power. He had a point, at least in theory. But as we'll see in the next chapter, traditional economics can't predict its way into tomorrow's inflation rate. It's notorious for being unable to predict macro- or microeconomic behavior. Without predictive power (or anything under the hood), it's merely a chimera.

Scientific theories are open to empirical testing. They don't just make predictions; experiments can prove them wrong. That's how science advances. But as we saw in the previous section, there is little that throws the free marketers off their game. Massive economic collapse of developing countries under the Washington Consensus? Market wasn't free enough. Huge financial collapse when Wall Street got to play by its own rules? Freddie Mac and Fannie May distorted the market. If any failure of a theory results in declaring the experiment invalid, the theory is worthless.

The "free as you can make it" argument is a fiction generated by the model, not a lesson for the real world. Even Alan Greenspan, champion of the neoconservative movement and acolyte of Ayn Rand, admitted in 1984 (in a remark whose irony can only be noted with a heavy dose of regret), "A surprising problem is that a number of economists are not able to distinguish between the economic models we construct and the real world."[69] Greenspan, as the Oracle of Washington, went on to lecture the world about free-market economics even while the sub-prime crisis swallowed Wall Street whole.

Frictionless surfaces in physics can draw out basic principles at play in the complex, messy real world, and there are useful lessons we can draw from economic models. A softer lesson, one that advocates a limited sort of freedom to operate in the market, or emphasizes free as one of many properties to be balanced (like stability, sustainability, and equality), is more reasonable. But taken to an extreme, which is exactly what the neoconservative myth about the sanctity of the free market does, the model is both unhelpful and untrue.

It's sometimes said equilibrium applies over the long term and that instability and inconsistency are just bumps on the road. Short-term turbulence is just the system seeking equilibrium, which is more like a guiding star than a current state of affairs. British economist John Maynard Keynes saw that criticism coming years ago: "But this long run is a misleading guide to current affairs. In the long run we are all dead. Economists set themselves too easy, too useless a task if in tempestuous seasons they can only tell us that when the storm is long past the sea is flat again."[70]

Free market fundamentalism's core idea — that free markets bring equilibrium and the best of all possible worlds — is outdated; it's based on ancient mathematics. As Eric Beinhocker, author of *The Origin of Wealth*, puts it, "Many of the 'big ideas' of the field are now well over a century old, and too many of the field's formal theories and mathematical models are either hamstrung by unrealistic assumptions or directly contradicted by real-world data."[71] Prior to the financial meltdown of 2008, even Greenspan said, "we really do not know how [the economy] works . . . the old models are just not working."[72]

It's time for some fresh thinking.

Dynamic Systems and Evolutionary Economics

Equilibrium is not a helpful model for the economy. It's lifeless, simple, devoid of movement. It exists only in test tubes, balls sitting quietly at the bottom of bowls, and empty space. It's a dead end. Almost anything of interest we see around us operates far from equilibrium, using external sources of energy to maintain the integrity of some dynamic structure. That's precisely what makes it interesting! The turbulent flows of a river, ice crystals forming, hurricanes, the waves of an ocean, and all of life itself, from individual cells to brains to online communities and entire ecosystems, these are all dynamic systems. None are in equilibrium. Neither is the global economy.

The core idea behind dynamic systems is simple: large-scale, complex stuff (like water flows or human minds) are patterns that emerge from lots of simple, smaller-scale stuff (like water molecules or neurons). Dynamic systems emerge as a dance of material. The patterns evolve over time. Energy keeps it all going. Thoughts emerge when lots of neurons fire in the right pattern. Whirlpools and eddies emerge when water molecules move the right way. We describe these phenomena with the modern language of dynamic (or complex) systems theory.[73]

A relative newcomer to economics—evolutionary economics[74]—provides a view of the modern economy as the mother of all emergent phenomena: it's the result of billions of people doing trillions of transactions, day in and day out.

An example like the computer program Game of Life is a good place to start in understanding dynamic systems. Consider a grid of squares that can be randomly filled with dots. Some squares are filled (alive) and some are empty (dead). Collections of dots form patterns—like blobs—on the board. The system evolves, step by step, according to a simple set of rules. For example, each dot surrounded by five or more dots dies but if it's surrounded by four or less, it lives, or a dead square surrounded by exactly five dots comes to life and so on. Successive iterations generate a new board. Speed it up on a computer and you can see the system evolve as blobs change shape. You can add layers of complexity; for example, the rules themselves can change depending on the total number of dots.

The board sometimes displays uninteresting, random behavior. It can also sink into one stable, unchanging pattern—an equilibrium. But occasionally, depending on the rules and the initial scattering of dots, patterns that are complex, even astounding, can emerge. Stable shapes called flyers spin off from giant undulating blobs, moving as coherent patterns across the screen. Flyers hit other blobs, causing new flyers to spin off. Sometimes unstable bits try to fly from the collision but peter out. Blobs morph between two stable patterns, form never-ending new shapes, or stay still. Blobs and flyers are like emerging dynamic systems, patterns of behavior spanning large numbers of dots that demonstrate complexity and stability at a scale far removed from the rules themselves. The lesson is that simple, iterative microscale rules can give rise to complex macroscale behavior. Think of the dots as molecules, neurons, or cells and the rules as the physics of interaction.

Now extend this toy example so it's a metaphor for the global economy. It begins to look more like an evolving web of increasing complexity than a system in equilibrium. Think of the dots as billions of people exchanging money and goods, making a living, spending, and earning. The rules are market conditions that govern transactions like contract law, interest rates, tariffs, electronic financial markets, and so on. Emerging patterns of money flow can be thought of as corporations, competitive advantages, markets, products or services, whatever. Maybe the flyers are new inventions. Money flows can measure many things.

Successive iterations generate new relationships and possibilities. Emerging financial and technological innovations can change the game itself, opening up previously unforeseen possibilities: new goods to trade, new ways of trading those goods, and new ways to make a living. Human creativity drives ever-increasing complexity in the evolving global economic web. We change the rules of the game as we play.

Take the role of technology. A new invention can make other inventions possible, as well as make entirely new ways of making a living possible. Metal casting made possible (among other things) the internal combustion engine, which opened up the possibility of cars, which made room for traffic lights, the Ford factory, car leasing companies, the drive-in restaurant, the great cross-country road trip, etc. There's an order of operations at play. The transistor had to precede the microchip, which in turn preceded the computer, the Internet, our highly integrated global telecommunications and data systems.

When someone invents something or applies technology in new ways, it opens space for more things to get invented. Possibilities rise exponentially. Predicting future states in any detail is impossible because you need know in advance all the ways each existing piece of technology might be applied and what new ones might arise. Stuart Kauffman, founding director of the Institute for Biocomplexity and

Informatics at the University of Calgary and luminary at the Santa Fe Institute, calls the set of possible next steps the "adjacent possible." As the system gets more complex, the adjacent possible grows. Complexity builds on itself.

But inventions can do more than enable other inventions, they can change the most fundamental rules of the market. Software turned the old law of diminishing returns on its head. It introduced a new law: the law of *increasing* returns. Normally, incremental consumption reduces the value of something. The next apple is worth a little less than the previous apple because people are getting sick of apples. Incremental consumption of the Windows operating system increases its value because it increases the possibility that it will become an industry standard. The next user of Facebook is of more value than the previous user because value of any one user depends on the total number of users on the system.

Over the past several thousand years, the global economy has grown from a few hundred goods to an estimated ten billion. It's humanity's most complex creation. Dynamic systems theory takes seriously the idea that the properties of the system itself change along with the possible directions in which the system can next evolve. That's partly what makes the system impossible to predict and keeps it far from equilibrium. Even those in the inventing game can't see what's ahead. Shortly after the computer was invented, the president of IBM, Thomas Watson, Sr., didn't think the world would need more than a couple dozen IBM mainframe computers. Bill Gates thought a few kilobytes of memory was enough!

But technology's effect cuts two ways: it opens new pathways but simultaneously cuts others off. Once King Coal's competitive advantages — it's cheap, ubiquitous, and reliable — got it established, coal became entrenched. Hundreds of billions of dollars were spent on infrastructure. Long-term contracts were signed. Investors expected returns. Market forces generated vested interests, oligopolies, established practices, and lobbyists. By the time we realized carbon dioxide could bake us, coal's interconnectedness had locked it in. Even when renewables are cheaper, that's not enough. Coal builds pathways in the economic web that block innovation. Plants with twenty-year contracts for power and built to run fifty years won't get shut off without confrontation or compensation. For renewables to compete, we must innovate across technology, communications, finance, politics, and economics.

Traditional economics incorporates technological change, but not in the same way. Most famously, Austrian-American economist Joseph Schumpeter acknowledged technology's capacity to cause massive change through "Schumpeterian gales of creative destruction." These destructive forces are often used (sometimes legitimately) to defend the harsh vagaries of the market, but change remains an external force applied *to* the system rather than being a key component *of* the system. That's a key difference — like treating the pain of a fever rather than the fever itself. That's something new and important about dynamic systems.

Technology is only one way the economic web modifies itself. There are others. Financial innovation enables new ways of doing business. The invention of the initial public offering (IPO), debt capital, central banking, collateralized debt obligations, and all the rest of the modern market economy, each of these innovations changes market rules with new interactive tools. Social and political innovations bring their own layers of complexity.

There is nothing like equilibrium here. Innovation sweeps away old ways of doing things, opens up new ways of making a living, creates new possibilities, and changes the rules of the game. Innovation is what makes our dynamic, evolving economic system fundamentally creative and unpredictable.

That doesn't mean we cannot have confidence in the future. Complexity is not randomness but a rich mix of order and chaos. Weather systems are complex. We can't predict Toronto's weather in June 2025, but we can be confident that the average temperature that month will be higher than in December of the same year. Coherent patterns remain. Within the complexity we pick out order: "the economy does evolve in ways that look predictable."[75] Rough rules still apply: economies of scale bring down manufacturing costs, shortages of supply choke mass production, scarce resources will run out, and learning curves confer lasting advantage.

The trick is to figure out which long-term trends are fundamental and which are subject to change. As a venture capitalist, I'm confident the world will de-carbon its economy. What I can't be sure of are the timing, the winning technologies, or which companies and jurisdictions will win the race.

This extended metaphor illustrates the fundamentals of dynamic systems like the global economy. They are complex: the interesting behavior sits somewhere between stable and random and on the boundary of chaos and order. Their patterns are emergent: things of interest are aggregations of huge numbers of smaller elements. Their behavior can't be predicted in detail: to see the patterns you have to let the game run. They are time-dependent: only over time is the interesting behavior visible. They are self-modifying: the system acts on itself, one state feeds into the next, possibly changing the underlying rules. They are path dependent: current possibilities are defined by the past, and the order in which things happen matters.

The new view of evolutionary economics is very different from the old one. In the new view, the global economy has unpredictable but stable patterns. In the old, it reliably trundles toward a single point of equilibrium. The goal of traditional economics was to increase predictability of macroeconomic behavior, but the new view emphasizes complexity over predictability. In the old view we are in a sort of eternal present, whereas in the new time is of the essence—the future is unknown but history matters. In the old view the economy is something static seeking equilibrium, and in the new it is dynamic, self-modifying, and endlessly creative. The old view has equilibrium, and the new view has a trajectory.

Recall that the traditional view argued that the equilibrium found by a free market necessarily brings about the best of all possible worlds, maximizing everyone's happiness. There's only one point of equilibrium. But a dynamic, evolving economy is highly path dependent—what came before defines what can happen now. So even if a long-term equilibrium were possible (which I doubt), there must be many different possible kinds of equilibrium. There is no guarantee we automatically float toward economic utopia. Bad choices lock in certain pathways—like that of high-carbon growth—that can take us to the *worst* of all possible worlds, or anything in between. We must not let a preference for "free" leave our market open to evolve without oversight. One thing we can do is constrain the direction in which our economy evolves.

WAKING THE FROG: CHOOSING OUR DIRECTION

In the dynamic systems view there is no equilibrium, but there is a direction. Instead of seeking a mythical equilibrium by minimizing rules, we influence the direction our economy takes by shaping those rules. We can't predict details of global spending patterns that emerge, but we can influence the general behavior. Contrary to traditional economics, the form of the rules matter. Where we end up decades hence depends largely on the rules of the game and the choices we make about our overarching goals. We can use lessons of the dynamic systems view of economics to make better policy choices about how to get off carbon-intensive energy.

Market rules affect broad economic outcomes, often in predictable ways. When liquidity of international capital flows was increased in the 1980s, for example, the expected higher levels of capital flow also increased the volatility of local currencies. There can also be unintended consequences. The derivatives that were to bring stability and decreased risk to financial markets did the opposite, largely because they were opaque and unregulated. Europe is finding that a centralized currency with decentralized politics brings huge local instabilities to sovereign debt markets. It is no surprise that companies choose to externalize costs, such as carbon dioxide, if they can. In each of these cases, the rules of the game influence the general direction the economy evolves, even if details remain complex.

Four useful principles emerge from the dynamic systems view of the economy.

First, the market is unmatched in its creative potential. There's nothing like it. Nobody could design the patterns of macroeconomic behavior that emerge from trillions of daily transactions. Adam Smith's invisible hand is real. We do well to emphasize the market's creative potential and ability to efficiently coordinate global economic behavior. On this point, the neocons and I are not far apart: the market is champion. It's like a powerful evolutionary search mechanism, seamlessly rooting out winners and losers, allocating resources, and generating complexity. The question is "for what is it searching?" In the absence of an overarching goal, it's best left alone.

Second, we do have overarching goals toward which we can aim. For good reason we worry about economic equality, injustice, environmental damage, global peace, political stability, and so on. The market by itself may or may not achieve any of those goals. The jury is out on all of them, but it's certain a freely evolving market will not solve climate disruption. Left to its own devices, it will effectively, creatively, and quickly burn every last kilo of economically viable fuel in the ground. The dodo went extinct. We don't want to be the dodo. We can guide the market toward our goals because we invented the market.

Third, we cannot effectively pick technological winners and losers independent of the market. The old leftist notion of government-imposed solutions is as outdated a notion as equilibrium. If the economy is too complex to model, how on earth can a centralized government pick effective solutions? Only the market, evolving in real-time and in real-world conditions, can effectively seek out winning technologies and financial models.

Fourth, early actions matter because the evolving economy is path dependent. We've seen how changing technology opens new pathways down which the economy might evolve, but embedded infrastructure and vested interests do the opposite, limiting new pathways and entrenching the status quo. Infrastructure lasts a long time and provides the context into which emerging technologies evolve. An electrical network designed for large centralized power plants, for example, inhibits distributed production technologies.

We can apply these principles and intelligently set market rules to solve climate disruption: the market's unmatched creative force can be effectively steered toward a low-carbon economy by changing the basic rules of the game. We cannot predict in advance what form solutions will take, but in acting quickly we ensure there is less chance of locking in inferior technology. We lose the leftist hubris of thinking we can decide in advance what form the solution takes but simultaneously dismiss the right's claim of the logical priority of freedom over market.

Even Milton Friedman acknowledged the importance of the rules of the game: "The existence of a free market does not of course eliminate the need for government. On the contrary, government is essential both as a forum for determining the 'rules of the game' and as an umpire to interpret and enforce the rules decided on."[76] All I've added is the importance of acknowledging that the rules can incorporate an important overarching, long-term goal—like a low-carbon economy that maintains a functioning global ecosystem.

Think of the rules of the game as wires in a vineyard: they guide the natural growth of the plant in order to maximize something (in the case of a vineyard, wine production). Another approach is to think of carbon as a problem to solve and the rules as the search mechanism for the most effective solution. The carbon-solving rules should be as basic as possible, affect the maximum number of transactions, and provide a general direction without dictating the details. In other words, we need a price on carbon.

Suggested Reading

Eric D. Beinhocker, *The Origin of Wealth: Evolution, Complexity, and the Radical Remaking of Economics* (2006)

Stuart Kauffman, *Reinventing the Sacred: A New View of Science, Reason, and Religion* (2008)

Michael Lind, *Land of Promise: An Economic History of the United States* (2012)

Joseph E. Stiglitz, *Globalization and Its Discontents* (2003)

CHAPTER 4

Economics: The Dismal Science

"What's the point building complex scientific models if you're going to plug them into dumb economic models?"
— Lee Smolin, physicist (said during a private conversation)

"The central cause of [economics'] failure was the desire for an all-encompassing, intellectually elegant approach that gave economists a chance to show off their mathematical prowess."
— Paul Krugman, economist

In the summer of 2011, at an international conference on district energy systems in Toronto, I had the opportunity to debate Danish statistician Bjorn Lomborg, also the author of climate quietist best-sellers *The Sceptical Environmentalist* and *Cool It*. Lomborg is a disarmingly charming speaker and a notoriously slippery debater. He has long been the darling of the "do-nothing" crowd, providing comfort to those who prefer not to worry about climate disruption. Debating Lomborg was something I looked forward to with great interest.

Lomborg might be a highly controversial[77] figure within the science community, but he has emerged as a champion for inaction on climate disruption in popular culture and political circles. He speaks in the calm, rational voice of economics. His language is that of cost-benefit analyses, discount rates, and investment criteria. His position is optimistic. A clear and consistent advocate of a slow and steady response to climate disruption, his message is simple: our best bet is to take it easy, relax, and just let climate disruption happen — and he's got the facts and figures to prove it.

While he now admits climate disruption is a "bit of a problem," Lomborg argues it's not the singularly dangerous threat to our growth — even our existence — that commentators such as David Suzuki, Nicholas Stern, Al Gore, and I would have it. Using the seemingly irrefutable language of dollars and cents, he argues that spending big money to slow emissions of greenhouse gases is a bad investment. Sure, we could spend a bit of money on research, but we're best off ignoring the problem for now and saving the heavy lifting for later.

If this were just about Lomborg, it wouldn't be worth worrying about. The 2010 movie version of *Cool It* was a flop, and he's been under investigation several times by the professional science community. But his message is seductive, shared by many, and dominates public discussion about the costs of acting on carbon. It provides cover for those who would put off what is admittedly the hard work of reducing carbon emissions. With a few brave exceptions, like the *Stern Review on the Economics of Climate Change*, Lomborg's "slow and steady" message reflects the echo chamber that is mainstream economic commentary.

We need economic analysis, of course. The cold, hard language of costs and benefits permeates our punditry and politics, informs captains of industry, and remains central to public policy. It would be silly to plan policy without good economic modeling. It's important to know what you're getting into, what the benefits are, and what costs are.

But intuitively, it seemed to me that Lomborg's arguments *must* be flawed since their conclusions are absurd. Don't worry about climate disruption? It's really too expensive to start retrofitting our buildings? Intuition is no way to debate an economist, of course, and the appropriate response has to be an analytic argument in the same language. Here was a chance to get to the bottom of what seemed to be a deep mystery: how so many people—trained at top universities in economics, statistics, and social sciences—can come to the same absurd conclusion.

Instead of just accepting his barrage of numbers, I decided to find out where all those numbers came from. Like any good philosopher, I went to the source. What I found was alarming. It's not just that the models generating these numbers are wrong; they're glaringly, admittedly, and openly wrong. The figures that drive the analysis are complete nonsense, dressed up to look legitimate.

The economic model behind a lot of climate-related economic analysis is called the Dynamic Integrated model of Climate and the Economy, or DICE. DICE is typical of a genre of models that all make the same two fundamental mistakes. First, it simplifies the climate science so much that it becomes nearly useless—a simplification its authors freely admit. Broken models generate bad data. Worse, these models churn out bad data with breathtakingly unwarranted precision. Lots of decimal places induce false confidence in the results.

Equally important is criticism that's been aimed at economics for some time: in attempting to develop a language that is "values-free," and thus gain the status afforded a quantitative science, economics impoverishes itself to such an extent that it becomes incapable of serious commentary on much of the complex human world. It's for this reason economics has sometimes been called the dismal science. Economics does itself a disservice when it tries to generate all the answers within the self-imposed boundaries of a wholly quantitative and empirical discipline.

The world is too complex to submit to our predictive power—history is littered with unexpected outcomes and failed predictions. Numbers can't capture the full complexity of human behavior, including questions of ethics and intergenerational justice. The good news is there are new and more collaborative ways to think about the economics of climate disruption. It doesn't have to remain such a dismal science.

Rolling the DICE

Look to the heart of any economic analysis of climate disruption and you'll find a cost-benefit analysis (CBA). Lomborg draws from the classical cost-benefit model developed by Yale economist William Nordhaus, the primary author of DICE. DICE calculates the net value of a climate disruption policy by comparing the costs of implementation — carbon taxes, the compliance costs to businesses, and so on — with the benefits brought by avoiding environmental damage. These costs and benefits occur over a long period of time, and so the comparison is made easier with a discount rate that yields a net present value.[78]

Lomborg uses DICE to argue that "[e]ven if all countries lived up to their [Kyoto] commitments . . . the expected temperature increase of 4.7°F [2.6°C] could be postponed just five years, from 2100 to 2105 . . . [and] if no other treaty replaces Kyoto after 2012, its total effect will have been to postpone the rise in global temperatures a bit less than seven days in 2100."[79] Predicting a rise in temperature is the start of any (climate-related) CBA. A policy reduces carbon emissions, which over time lowers total carbon in the atmosphere. From there, we calculate the expected temperature reduction. It turns out Kyoto won't do much by itself — a point with which I fully agree.

So far, so good, but notice Lomborg isn't just saying Kyoto won't do much; he's saying it would delay a rise in global temperatures by less than a week by the end of the century. His is not a qualitative statement ("won't do much") but a quantified one ("less than a week by century's end"). He shows great confidence in the predictive ability of his model. Seven days over a century implies an accuracy of less than 0.02 percent.

The next step is to calculate the cost of implementing a policy like Kyoto,[80] for which estimates vary wildly (as you might expect). One can imagine what we might do to lower emissions: retrofit buildings;

retool the auto fleet to be more efficient; replace gasoline with biofuels; build solar, hydro, wind, and geothermal plants to replace coal; and so on. One way to do this is to put a price on carbon — arguably the most powerful tool a market economy has at its disposal. To get the total cost, you just add up all those taxes (or costs of regulations and so on). Lomborg's own numbers indicate that the "total cost over the coming century turns out to be more than $5 trillion."[81]

So we have a plan and its cost. The benefits are more complicated to compute, since we first have to take into account a vast array of climate impacts on elements such as "agriculture, forestry, fisheries, energy, water supplies, infrastructure, hurricanes, drought damage, coastal protection, land loss . . . wetlands, human and animal survival, pollution."[82] The policy's benefits are a measure of the avoided impacts over time. It's tricky, but DICE faithfully cranks out a number: "The total benefit for the world comes in at $2 trillion."[83] Surely Kyoto's a bad idea: $5 trillion in spending yields just $2 trillion in benefits!

Lomborg knows a bad deal when he sees one. For him, Kyoto is a con that stretches far into the future: "It turns out that for the first 170 years the costs are greater than the benefits . . . [and] the total benefits outweigh the total costs, around 2250."[84] Apparently the first generation to get any benefit won't be born until sometime in the twenty-fourth century. Wow! Four hundred years of overpaying. Sounds worse than a bank bailout.

If Kyoto's a bad deal, how do we find a good one? That's simple: you find the "optimal path" by juggling the numbers a bit. Tune the knob on tax rates to get a new cost, calculate the associated drop in carbon levels and temperature rise, and that gives you a measure of the benefits. Keep tuning until the benefits outweigh the costs. Out pops the optimum rate of both investment and warming.

The net result? Slow and steady is the way to go. The optimum path is to let the Earth warm by exactly 2.6°C (4.7°F) by century's end.

Start with a bit of money for R&D, but save the real work for when we find the magic bullet that makes fossil fuels obsolete. This is a comforting result because it tells us we are justified in sitting on our butts and not doing very much.

But it's wrong.

Where would DICE take us? "We should cut CO_2 by more than what the emasculated Kyoto will manage," Lomborg says, "but still only by 5 percent, moving to 10 percent by the end of the century."[85] This brings atmospheric carbon levels to somewhere between 700 and 800 parts per million (ppm) by the end of the twenty-first century. This view is representative of mainstream economic analyses: "All peer-reviewed economic analyses show we should cut CO_2 only moderately."[86]

Yet, no one with any serious knowledge of climate science thinks 800 ppm is a place modern civilization can go. The planet may make it through that level of carbon in one piece, but our civilization won't. Simply put, DICE takes us to a very high probability of global chaos. It's a clear example of garbage in (bad science), garbage out (bad advice). Yet DICE remains the consensus view among mainstream economists. There is a childhood tale that teaches us how badly groupthink can go—it's called *The Emperor Has No Clothes*.

It seems to me (and others) that Lomborg can't have any idea what these levels of carbon really mean, and a lot of his passing comments appear designed to make people who worry about climate disruption feel a bit silly: "We may notice people wearing fewer layers of clothes on a winter's evening."[87] That sounds nice—maybe I can throw out my winter coat! He compares global average warming to the heat island effect created by warm summertime asphalt in large cities: "New York has a . . . night-time urban heat island of 7°F [3.9°C]. . . . Many of these increases . . . are the same or bigger than the 4.7°F [2.6°C] that we expect to see over the coming full century."[88] New Yorkers have clearly adapted to that level of warming using air conditioners, so how bad can it be?

These may be rhetorically effective gimmicks that play well to an adoring crowd, but they are easily dismissed displays of statistical illiteracy.[89] Comparing New York's heat island effect with a rise in the global average temperature is patent nonsense. It's flippant and irresponsible to imply we can air-condition our way out of climate disruption.

A rise in global average temperatures associated with 800 ppm of atmospheric carbon brings systemic risks throughout our food supply with massive increased risks of droughts, severe weather, and flooding in coastal areas. By the end of the century, those risks are off the charts! Resource scarcity leads to wars over food and water. Our ever-aging infrastructure will be at risk of being consistently overwhelmed by storms, water, and fire—and all the attendant physical and financial misery. Ocean levels will eventually rise not by meters but by a hundred meters (328 feet) or more because the ice caps will completely melt over time. That kind of carbon level is a "bit of a problem"; it's the end of comfortable life as most of us know it now!

This isn't about Lomborg. He's just the poster child for the "do nothing, don't worry" crowd. Something akin to that view dominates mainstream economic thinking and is considered a very robust result. Nordhaus, writing in a mainstream policy-wonk paper, says, "Current assessments determine that the 'optimal policy' calls for a relatively modest level of control of CO_2."[90] This nonsense informs policy-makers, makes those who advocate for strong action look like worrywarts, and kneecaps any captains of industry brave enough to stand up and demand real action.

I emphasize there are some basic points on which Lomborg and I agree. Kyoto is a drop in the bucket. There will be a cost associated with implementing the agreement, and that cost will certainly increase if we go past Kyoto (as we must). The benefits of action occur mainly in the future. But that's the limit of agreement. My conclusions are the opposite of Lomborg's and those of his cronies: I think we should press the accelerator and spend as much money as we can afford on reducing emissions in the short term.

Why such a different point of view if we agree on some basics? Because DICE is deeply flawed. It's got the science wrong in fundamental ways. It ignores the biggest risks and eliminates scientific uncertainty instead of dealing with it honestly. An unwarranted level of precision confers unjustified confidence in its ability to predict the future. And, most importantly, the outcome is rigged from the start by cherry-picking highly influential variables, in particular, the discount rate.

Complex Science, Simple Economics, and Potential Catastrophes

The primary problem with DICE is that it dumbs down the massive complexity of climate science to a few simple linear equations, in the process ignoring the non-linear behavior that has scientists most worried.[91] That makes sense if your goal is to cram the science into a spreadsheet,[92] but the simplicity comes at a cost. It's like trying to predict someone's behavior based on the shoes they're wearing or stock-market futures based on weather in Dallas: you simply don't have enough information. Here are just a few of the complex behaviors that keep scientists up at night, but that DICE ignores.

We now know sudden, catastrophic changes in climate are a real possibility. These low-probability but highly consequential events include the breakdown of ocean circulation patterns and the melting of polar ice caps. Neither the precise timing of these catastrophic events, nor their effects, can be treated with any degree of certainty, but we do know they're worth worrying about. DICE ignores the possibility of any nasty surprises.

We also know the climate system has lots of positive feedbacks. Positive feedback is like the squeal of a microphone held to a speaker. The noise gets amplified, which causes more noise, which in turn gets amplified and so on. In the climate system, greenhouse gases cause warming, which releases more greenhouse gases and so on. Carbon sinks can become sources: warming oceans and soils release stored carbon. Dying forests do the same. The melting north already belches huge amounts of methane.[93] Warming deep oceans might follow suit, releasing vast amounts of the potent greenhouse gas currently locked up inside a lattice of ice (called clathrate hydrates). Ice in the north reflects heat; as it melts, the darker ocean absorbs more heat, and so on. As these triggers go off, we could be stuck watching helplessly as the climate shifts all by itself into another, very hostile, equilibrium. DICE ignores positive feedbacks.

It might seem cheaper to invest in R&D and wait for the cost of clean technology to come down, but we don't have that luxury. These triggers release long before we feel their effects, so an aggressive early response is essential to reduce the probability of hitting the triggers. Timing matters. In technical terms, the climate is path dependent. The path we choose at the start determines what choices we have later on. DICE ignores path dependency.

Variability in extreme weather events such as Category 4 or 5 hurricanes, floods, and droughts are not well understood. There's not enough history of warming to generate empirical evidence. Paradoxically, waiting around for sound historical trends on extreme weather events means we've waited too long to use the data. It's like waiting for the horses to bolt out of the barn and gallop over the hill before we analyze the benefits of a good door. But we do understand the underlying theory of these extreme weather events well enough to have confidence that there is significantly increased risk. DICE ignores that risk.

DICE is linear and simple, but climate science is non-linear and complex. DICE treats the climate system like a basic oven, with temperatures being precisely controlled by policy and technology, but the climate is nothing like an oven. It is more like a large, sleepy dog we're poking with a stick.[94] We don't know quite when the dog will wake up, growl, or leap at us, but we know it will, eventually. Every time we build a coal plant, we poke that dog. Every time we build another pipeline from the Canadian tar sands[95] or sell an exploratory license to drill for oil in the Arctic, we poke that dog. Mainstream economic models such as DICE cannot provide good advice when they oversimplify the science.

These criticisms of DICE are not controversial. Nordhaus himself acknowledges that DICE doesn't entertain the possibility of low-probability, highly catastrophic events: "This approach assumes[96] that there are no genuinely catastrophic outcomes that would wipe out the human species or destroy the fabric of human civilizations."[97] An honest academic, Nordhaus advises us to fill this hole: "Estimating the likelihood of, and dealing with, potentially catastrophic outcomes is one of the continuing important subjects of research for the natural and social sciences."[98] Indeed. While Rome burns, he notes the need to study the flammability of buildings.

It should come as no surprise that Lomborg is not worried about catastrophic climate disruption: the model he uses assumes it can't happen. Nordhaus cautions us on relying on just one model: "No sensible policy-maker would base the globe's future on a single model [or] a single set of computer runs."[99] Lomborg shows no such caution.

We have learned the hard way what happens when we ignore risk and uncertainty. For years, financial analysts ignored the risk of subprime mortgages because their models didn't include the possibility of house prices falling everywhere all at once. In 2008, that's exactly what happened, and the U.S. and global financial systems ground to a halt as the global financial crisis spread. Japan recently learned the hard way what happens when you ignore the uncertainty of low-probability but catastrophic events when the devastating 2011 Tohoku earthquake and subsequent tsunami hit. Japan is well on its way to recovery, and governments around the world rescued the financial system with trillions in bailouts and backstops. There will be no such rescue if the climate hits a tipping point.

Cascades of Uncertainty

"There are known knowns. These are things we know that we know. There are known unknowns. That is to say, there are things that we know we don't know. But there are also unknown unknowns. There are things we don't know we don't know."

— U.S. secretary of defense Donald Rumsfeld, 2006

Risk is a term we use to describe known probabilities. There's a risk I'll roll a seven playing craps or my poker opponent will get a flush on the river card. I don't know the result in advance, but I understand enough about the game to know the probability. Uncertainty is different. It's an unknown probability. There is some uncertainty about whether the Boston Celtics will beat the New York Knicks. Bookies give odds that reflect risk, but there remains a degree of uncertainty about the outcome. There is great uncertainty about whether China's rice paddies will dry out as carbon counts climb, and we have no real idea of the probability. Uncertainty is a risk we don't fully understand.

Risks are like Donald Rumsfeld's "known knowns." We have full statistical understanding of risk even when we can't predict the event. Dice rolls are a risk. Uncertainty is like the "known unknown." We know what it's about even if our knowledge is imperfect. In the rice paddy example, we know the event (they dry out) but not the probability of it happening. Rumsfeld's "unknown unknown" goes one step further. It's really just a way of saying we have no idea what's around the next corner. Unknown unknowns are what make life so gloriously interesting! But as the world found out on September 11, 2001, and later during Rumsfeld's war in Iraq, unknown unknowns are not always benign. Their existence should make us cautious about predicting the future.

Uncertainty cascades through DICE. Each step in the model's long causal chain introduces a different kind of uncertainty. For some, we have a pretty good idea of the probability of a particular effect given a cause, say, from emission levels through the carbon cycle. For others, we have less confidence, say, from carbon levels to global average warming or from that to localized climate effects. For a few, we have to admit we have very little confidence at all. What will be the effect of a rise in temperature on global fisheries? On the economy as a whole? As you multiply these effects together, the uncertainty amplifies. Uncertainty is generated by both science and economics. Finally, because a human-warmed Earth is a first-of-a-kind experiment, we have to admit it will likely bring a bunch of unknown unknowns to the table.

Critics of action often point to the deep uncertainty in climate models as a reason for inaction. We are urged to wait for more information. But some uncertainty is inherent in complex systems. You just can't make it go away. That's why we can't predict the stock market, always give the right odds for the Celtics, or provide detailed daily weather information for next month. Uncertainty introduces unknown levels of risk. Our difficult task is to acknowledge and manage those risks and uncertainties. We normally do that by speaking in terms of probabilities, but even dealing in probabilities is difficult when it comes to catastrophic climate disruption because we have to make all sorts of unwarranted assumptions. It's messy, but only if we continue to think we need to quantify all this stuff.

Let's run a thought experiment. To what sort of risk does Lomborg's optimal solution expose us? Remember, his solution is to go slow and steady, letting carbon concentrations level off somewhere around 800 ppm. Lomborg claims this results in just 2.6°C (4.7°F) of warming, to which we'll presumably adapt by wearing fewer layers in winter and turning up the AC in summer.

Now let's define "catastrophic warming" as some definitive line in the sand that destroys the security of human civilization and causes untold suffering. Following U.S. economist Norman Weitzman,[100] whose current research is on environmental economics, particularly climate change, I'll pick a global average of 10°C (18°F) of warming as catastrophic. Why? Studies[101] show this level of warming makes most of our world unliveable, with a wet-bulb temperature[102] of 35°C (95°F) at least once a year, where "death from heat stress would ensue after about six hours of exposure."[103] This study minimizes actual suffering at a warming of 10°C (18°F), since it assumes the person in question is "out of the sun, in a gale-force wind, doused with water, wearing no clothing and not working."[104] In other words, the wet-bulb temperature relates to a measure of how long you can keep a resting body alive. The highest wet-bulb temperature in the world today is 30°C (86°F).

Clearly, global average warming of 10°C (18°F) is catastrophic. What are the odds Lomborg's "take it easy" attitude gets us there? And how does DICE decide what we should spend to avoid such a nightmare world?

Let's start with the mother of unknown unknowns: the current rise in carbon is a one-of-a-kind event, unprecedented in human and geological history. The last time carbon levels were at 800 ppm was during the Eocene epoch, about 50 million years ago. That's long before humans were a twinkle in evolution's eye. More importantly, as far as we can tell, the current rate of carbon increase has *never* happened before. It took around 100,000 years for carbon to reach these levels during the Eocene. We're doing it in a tiny fraction of that time, in just one hundred years. That's a first-time, planet-sized experiment we can't possibly understand, but one we know is dangerous. We have no idea what sorts of surprises might lie in store.

Analysis starts by estimating the "equilibrium climate sensitivity," a measure of the long-term increase in temperature we get from a doubling of pre-industrial carbon levels (280 ppm). From there, it's not too hard to predict temperature rises for any level of carbon. These estimates come in the form of a "best estimate" temperature and a probability distribution function (PDF) that calculates the likelihood of hitting other temperatures.[105] The PDF is a curve on a graph, a line that lets you read off the odds of straying from the best estimate or average. Think back to your school days, when you were graded on a bell curve. It's the same thing. The bell curve is a standard statistical tool that measures how likely something is to differ from the statistical average (like marks, people's height, or a stock price). The bell curve translates uncertainty into risk.

Ignoring the details, the takeaway is this: the PDF curve of climate sensitivity gives a best-guess temperature (3°C/5.4°F) for a doubling of atmospheric carbon and a way of calculating the odds of a range of other temperatures. The faster the curve drops off, the less likely we are to see more extreme temperatures. Past a certain point (the bound), such extreme temperatures are effectively considered impossible. So the curve captures the uncertainty in climate sensitivity. We extend the analysis in very simple ways to calculate the odds of hitting some temperature for any given carbon level.

At 800 ppm, the IPCC's standard analysis gives us a best estimate temperature rise of 4°C (7.2°F), with odds of going over 10°C (18°F) of just one-tenth of a percent, or one in a thousand. Four degrees is pretty bad, but at one in a thousand we don't really have to worry about the catastrophic world of 10°C (18°F) of warming, right? Wrong.

The curve the IPCC uses to generate those odds is arbitrary. Remember, there is no historical trend for this experiment. That means there's no reason to choose the bell curve over some other shape, other than it's a kind of statistical habit. The bell curve often works well in other disciplines, but it can also fail spectacularly. Scholar Nassim Taleb made famous the inadequacy of the bell curve in his best-selling book *The Black Swan*. A lot of interesting behavior, from financial markets to people's expectations, doesn't fit the bell curve. Just because it's a standard tool doesn't make it the *right* tool. We could just as easily pick other curves of many different shapes. This shows how an unknown unknown can affect real analysis.

Take, as an alternative, something called the Pareto curve. The main difference between it and the bell curve is that the bell drops off much faster.[106] The Pareto is said to have "fat tails," which means it's a more conservative choice since it will assign higher probabilities to more extreme events. What are the odds of hitting 10°C (18°F) of warming at 800 ppm with a standard Pareto curve? Nearly five percent.[107] So with a Pareto curve and Lomborg's recommendation, there's a one in twenty chance we'll bake the entire planet. Do you like those odds?

I have no idea which curve is appropriate, and neither does anyone else—perhaps it's something even more conservative than the Pareto. But choosing to minimize the probability of extreme events by fiat does not seem wise, given what's at stake. What are the real odds of a 10°C (18°F) world under Lomborg's plan? Is it DICE's one in a thousand? Is it the Pareto's one in twenty? Is it even higher? Nobody knows, but it certainly isn't negligible. We need better treatment of scientific uncertainty than an arbitrary choice of PDF curve by a math geek with a spreadsheet. That just hides it under the rug.

Don't forget that the relationship between carbon and temperature is just one step in a long chain of cause and effect (which itself includes a number of steps). PDFs get assigned each step of the way to capture the uncertainty behind that bit of the science. On each step we find a bell curve, mainly because uncertainty behaves so much better under a bell than a Pareto curve, not because we have any reason to think a bell curve is the right one to use. DICE tames uncertainty by filtering out the very events we should be most worried about.

It gets worse. DICE also vastly underestimates the damage these catastrophic events might cause, mainly because it hasn't figured out how to do so. It hasn't figured that out partly because it has assigned catastrophic warming such a low probability (so why bother?) and partly because it's hard to calculate the damage with the certainty of numbers with decimal places. There's a strange and vicious circle at work here.

The "damage function" in DICE defines the cost associated with some level of warming. It's an extremely simple formula that's meant to apply only to small amounts of warming.[108] Applying it to our 10°C (18°F) outcome generates some pretty ridiculous numbers.[109] Total economic damage at 10°C (18°F) would be less than a fifth of economic output. This means that when more than half of the planet is uninhabitable, we can't feed the current population, and the world is suffering unimagined social and political upheaval, including resource wars, more than four-fifths of economic output remains intact. Does this seem a reasonable damage estimate? Not a chance. And if asked directly, there aren't a lot of economists who think so, either.

And that's without taking into account assumed economic growth. Say it takes two hundred years to hit the 10°C (18°F) mark and growth is two percent per year between now and then. That means the one-fifth hit we can expect from entering this nightmare world can be translated into the equivalent effect of lowering compound growth from 2 to 1.9 percent. So the path to the catastrophic outcome of an uninhabitable planet translates to a tiny dent in our ever-growing economic pie.

When you put growth into the equation (which we always do), it's hard to imagine just what sort of catastrophic scenario DICE *would* have us take seriously.

A classic move in response to this sort of critique is to argue that as one resource is damaged or becomes scarce, it becomes economic to replace it with some other resource for which it can be substituted, given some tinkering. That's one reason the damage function in DICE is so linear and well behaved. As oil becomes scarce, we can replace it with biofuels. As wild fish become scarce, we can ramp up fish farms. As we run short on cotton, we can use polyester instead. But this is nonsense when we consider catastrophic climate disruption. You can't replace the wheat fields of Kansas with greenhouses in New Jersey. Substitutability only holds when the economic and ecological system as a whole remains robust. As we'll see in the next section, the fact that climate disruption is systemic, that it can affect everything all at once, changes the rules of the game.

Notoriously, climate scientists have a much higher estimate of the economic damage that would arise from catastrophic climate disruption. A recent study estimated the net present value of future climate damages (including cost of adaptation) runs into the hundreds of *trillions* of dollars.[110] This number is mind-boggling and really just says we're going to lose everything. Perhaps scientists have such a massive damage function because they are economically naive. Perhaps they're more realistic about what's coming our way. The truth is probably somewhere in the middle. Either way, it's prudent to acknowledge that the current damage function, imposed arbitrarily by economists, is hopelessly meagre.

What's the correct damage function for higher temperatures? Nobody knows, although there are lots of guesses. The full complexity of human social behavior, under duress in a stressed world, would have to be taken into account. No one knows how to figure this stuff out with precision. Surely, long before this happens, our concerns are not so much economic as moral and existential.

Worse, these last two steps—the probability function linking carbon to temperature and the damage function linking temperature to economic cost—add up to a one-two punch that's more than a simple sum of each. DICE estimates the cost of potential catastrophes by *multiplying* the probability and damage functions. DICE imposes an arbitrarily low value on both, without little empirical or theoretical support. Lowering both effectively eliminates climate catastrophe from the discussion. It's no wonder Lomborg can sleep at night; he's written the monsters out of the story.

Discounting the Future

Probability and damage functions rig the game. They make us feel safe when we're not, but the math can be a bit arcane and difficult to understand. If there is a single, simple variable that by itself can change the output of an economic model, it's the discount rate. Small variations in this well-understood global variable lead to massively different outcomes.[III] Tweak it and you can change everything. It's like bringing an NBA ringer to a local basketball game: it doesn't matter who else shows up; the outcome of the game is fixed before the opening buzzer.

A dollar in your hand today is worth more than a dollar in your hand a year from now. Why? Mainly because today's dollar can be invested so next year you have more than a dollar. At a five percent interest rate, you'd have a dollar and five cents. The discount rate is a percent—like an interest rate—that translates value from one time to another. It's how much you lower, per year, a future amount to put it in today's dollars. It's often set to the rate of return on capital investments.

There are lots of reasons to use a discount rate. For example, people are often risk averse, and so they value the certainty of having a dollar today more than they value the possibility of a dollar coming to them tomorrow, and they'll pay one cost today to avoid another, larger potential cost tomorrow. It also reflects the idea that we get richer over time, since the economy (generally speaking) grows. A dollar means less as we get wealthier. There are variations on this theme, but that's the main idea.

Reducing carbon emissions generates costs and benefits spread out over long periods, so it's difficult to compare them unless we apply the discount rate. It translates different values over time into a common currency—generally today's dollar. Just as we translate the cost of goods all over the world into a common currency—like the U.S. greenback—we can translate the cost of events that happen in the future into the common currency of today's money. A higher discount rate makes future climate damages look smaller; a lower discount rate does the opposite.

Here's why the discount rate is like an NBA ringer: just like the interest on your bank account, it's a compound rate, so it grows exponentially over time. Costs and benefits more than a few decades out are made irrelevant by small changes to the discount rate. Future disasters can seem innocuous if we subject them to a high enough discount rate. According to Weitzman, "The logic of compounding a constant positive interest rate forces us to say that what one might conceptualize as monumental—even earth-shaking—events like disastrous climate change, do not much matter when they occur in the deep future."[112] And since tomorrow's benefits occur later than today's costs, the benefits get hit harder than the costs. The discount rate rigs the outcome against climate action from the start.

An old parable illustrates the power of exponential growth. The inventor of chess so impressed the king with his new game, the king asked the inventor to suggest a reward for his work. The mathematically astute inventor asked the king for one grain of wheat for the first square of the board, two for the second, four for the third, and so on—doubling the number of grains for each square. The king thought this was a silly request and granted it. When the kingdom's treasurer tried to calculate the amount of wheat, he found there were not enough assets in the kingdom to cover the promise. In some versions of the story, the inventor becomes king and in others he is punished. The discount rate is another kind of exponential growth.

The DICE model uses a discount rate of four percent. Is that reasonable? A standard reply is to say four percent reflects existing market returns. In other words, it's an empirical measurement and not subject to interpretation. That may be true for the near future, say over the next decade or so, but does it make any sense for the distant future?

Let's say there's a sudden—and permanent—collapse of the entire global fishery in 2100 due to warming oceans.[113] By then, let's say the industry is roughly the same size as it is today in real terms—about $85 billion annually.[114] At a one percent discount rate, the lost income from a total collapse of that industry in 2100 is equivalent to $554 billion today.[115] This reflects the maximum cost we should bear (expressed in today's money but spread over the course of the century) to save that industry. In other words, it's worth spending a lot.

But at a four percent discount rate, its present value drops to $23 billion.[116] At eight percent, it's just $16 billion. That's all we should spend, over the entire century, to save the global fishing industry! Intuitively that makes no sense. The answer was rigged the moment we chose the discount rate. Given a high enough rate, it's hard to imagine what sort of damages a hundred years out we *could* take seriously.

DICE sets the discount rate equal to the existing market rate (what you'd earn on your money in a bank). That's standard fare in modern economics, since the market rate is meant to reflect actual preferences of consumers when it comes to risk aversion, the actual growth rate of the economy, and so on. So what's gone wrong? It's simple: there's no justification for setting the long-term discount rate for climate damages to today's market rate. It's just a habit economists have fallen into. There are lots of reasons why it's not justified. Here are just four of them.

First, the Industrial Revolution has only been around for two centuries. We have no historical precedent for thinking present-day rates of return apply over such lengthy periods of time. Lending contracts don't extend for centuries; they extend for maybe twenty or thirty years (think of the mortgage on your home). Markets reflect what individual people or companies do over the short and medium term; they do not reflect what society does over the very long term.

Second, why would interest rates have anything to say about consumer preferences about the ethical dilemma of leaving our grandkids with a beat-up planet? The market rate is set by large institutions, like central banks and corporations, and reflects their interests. There's no reason to think those rates have anything to do with how people might think about inflicting massive economic damage on future generations. The vast majority of people, and companies, are rate takers, not rate choosers. Even most governments have rates imposed upon them by their central banks, who are thinking about such things as inflation and currency devaluation.

There is an irreducibly ethical component to the discount rate as it's used in DICE. It's making decisions on how to impose economic costs on other generations. This is the ultimate conceit of economics: that long-term ethical values are expressed through short-term consumer behavior. It's ludicrous to think that consumers are somehow expressing those ethical values when they choose to buy government bonds or that the average rate of return of the stock market can shed some light on the difficult and fundamentally ethical questions of intergenerational justice.

Third, the rate of return varies massively, both between national economies and over time. Remember interest rates in the 1980s? In North America, they were in the high teens, and as a result many people lost their homes. Compare then to now, when the U.S. Treasury pays less than one percent. Compare Greek bonds in 2011 to U.S. bonds in the same year, and they differ by a factor of nearly twenty. It's true we can average these numbers out, but at any given time, they reflect current macroeconomic conditions, and those conditions change.

This brings us to the last, and most important, reason why using the market rate as the discount rate is a bad idea. Recall that the discount rate is meant to reflect the long-term growth rate of the global economy. Catastrophic climate disruption is exactly the sort of thing that will knock economic growth on its ass. That's the reason we're concerned in the first place: future generations might be poorer than us. The appropriate discount rate might be *negative*! A negative discount rate would flip DICE upside down.

The discount rate is normally used to analyze an investment — building a new bridge or hospital, investing in new equipment for a factory — and assumes the rest of the world remains constant. It's a global variable that's treated independently of the project under consideration. With catastrophic climate disruption, the project is not independent of the discount rate. They are interrelated, so the standard method of picking a discount rate and using it as an external, constant variable against which investments are judged makes no sense.

Picture a factory that makes T-shirts. The factory has a hole in the roof, and the owner must decide whether to fix the roof now or to wait until next year. With a high enough discount rate, he might decide to wait and invest the money in making more T-shirts to sell today. It's reasonable to take the chance a bit of rain might ruin a machine or two. But if a storm hits and floods the entire factory, there'll be no money next year to fix the roof. There would be no factory at all, and no way to recoup the money.

Climate disruption is like that storm: it could wreck our entire "factory." It represents *systemic* risk. Traditional economic models like DICE do not know how to deal with systemic risk. Simple, habitual uses of the discount rate are not applicable to threats that affect everything all at once.

Setting a discount rate appropriate to catastrophic climate disruption is not something you do by looking at the current market rate. It's not an empirical choice. It's not even a purely economic choice. Selecting the appropriate discount rate is an ethical choice. It's the way we decide to treat future generations. It determines how we value their economic security. Economics alone cannot determine that ethical framework. Claims to the contrary are hubris and contribute to the view of economics as a dismal science.

The Dangerous Certainty of False Confidence

Cascades of uncertainty roll through DICE and are multiplied many times by assumptions hiding behind the arcane language of probability distributions, damage functions, and discount rates. Nordhaus himself admits that the "pace and extent of warming are highly uncertain — particularly beyond next few decades," warning that DICE's predictions beyond that should be viewed with suspicion. He adds that our knowledge of climate damage is "very meagre,"[117] and "providing reliable estimates of the damages from climate change over the long run has proven extremely difficult."[118] Nordhaus openly admits we can't trust the numbers past the next couple of decades, especially when it comes to the damages side of the ledger.

So how can Lomborg possibly justify his opening gambit about temperatures a hundred years from now? Remember, he assumed an accuracy of less than 0.02 percent! And what on earth is he doing making predictions about the twenty-fourth century? Nordhaus's warnings are easy to find by anyone who cares to look. Lomborg chose to ignore them, so his analysis is wrong. Its breathtakingly unsupported degree of precision has the feel of someone who spends too much time playing with spreadsheets and not enough studying the science.

Lomborg uses precision like a club. Again and again, he hits you over the head with mind-numbingly precise cost comparisons that show how much better of a deal we get by spending our money on something else: "Kyoto Protocol would reduce malaria risk by 0.2 percent in 80 years";[119] implementing Kyoto would bring "a decrease in hurricane damage of about half a percent [by 2050]";[120] "in Bangladesh the loss [of land area due to rising oceans] will . . . be virtually nil at 0.000034 percent [by 2100]."[121] The list goes on and on. All these comparisons are invalid because his climate calculations are nonsense.

To take just one example, the uncertainty associated with the rise in ocean levels by the year 2100 is massive and depends on a boatload of factors, such as what tipping points are reached, how much of Greenland melts, our carbon emissions, and so on. The ocean rise by 2100 could be anywhere between half a meter to two meters (one to six feet). The potential variance from Lomborg's 0.000034 percent is many orders of magnitude. The same can be said of most of his predictions.

We learned in Grade 10 science class how uncertainty limits precision. If you multiply two numbers in a science experiment—say, the pressure and temperature of a gas in the classic equation $PV = nRT$—the degree of precision allowed in the final answer is limited by the less precise number going in. It's called significant digits. For example, 2 times 4.12 wouldn't get you 8.24. Rather, it would give you an answer of 8. That's because the decimal places after the 4 don't count. They're washed out by the implicit uncertainty contained in all the unknown decimals after the 2. The basic lesson is simple: the precision of a system is limited by its least precise component. Two points follow from this example. First, Lomborg doesn't provide any indication of the uncertainty contained in his numbers. Second, we often don't know their value within a factor of five or ten—so quoting any number with precesion is misleading. That Lomborg is a professional statistician yet fails to follow this elementary rule is alarming.

DICE is like a musket from the 1800s—great at close range but otherwise useless. Lomborg's claims are akin to taking out individual matchsticks time and time again on the other end of a football field using that musket. Lomborg's confidence, expressed in how many decimal places he uses, is unearned, dishonest, and highly misleading.

"Ah," I can hear the furrowed-browed team of economists say, "but we're doing the best we can. Missing details don't mean we're wrong—and it's certainly better than nothing. We've got to make policy choices on *some* basis. You can't just fly blind!" True enough. I'm not saying we go without good economic analysis, that's throwing the baby out with the bathwater, but we must acknowledge the limits of our models and build better ones that deal with the uncertainty in a more robust and open way.

As it stands, the DICE analysis does more harm than good. We're better off throwing it out when it comes to long-term planning. The false confidence that comes from meaningless precision provides unjustified support for inaction. It's steering us in precisely the wrong direction, with unearned authority.

FROM DISMAL SCIENCE TO COLLABORATIVE EXPERTISE

There's a reason economics is often called the dismal science. The insult is derived from the generally failed (but ongoing) effort of economists to fully capture the open-ended and complex nature of human economic behavior using only the tight, quantitative language of mathematics. Many of the humanities periodically go through stages best described as "science envy."[122] Psychologists, social scientists, and even philosophers have tried with varying degrees of success to capture the nuance of their language and theories using only empirical data and the rigor of math. Those efforts always fall short.

But economics, more often than any other social science, commits itself to the view that it's akin to physical science. Simply put, much of modern economics—particularly that which drives belief in free and unfettered markets—is purported to be capable of fully explaining market behavior (which includes human psychology and ethical preferences) without the need for qualitative, interdisciplinary discussion. This belief is partly driven by a desire to answer complex questions of policy and political priority in an authoritative language similar to that of science. High-level policy wonks prefer hard numbers and definite answers to foggy generalizations. Economists who provide these answers can benefit from a heightened professional stature.

Math is a necessary part of good economic theory, of course. Without math, economics would be greatly impoverished, and there is great benefit from the careful, measured analysis economics can provide. The danger lies in falling too much in love with the intellectual toys that provide the precision. It's one thing to give precise answers when possible, but it's quite another to bluster your way through deep complexity with fake precision. Proponents of DICE, and other similar economic models, have become the standard-bearers in economic analysis of climate disruption precisely because they provide hard numbers. But sometimes foggy generalizations are all we have, and we should admit it.

The complexity of our economic life cannot be quantified, predicted, or even fully described using only math. Thomas Sargent of

New York University who won the Nobel Prize in Economics in 2011 worked on something called "rational expectation theory."[123] The upshot is that the outcome of many economic situations depends on what people *expect* to happen. It's recursive, which means that expectations are dependent on some market condition, which in turn is affected by expectations and so on. At the macroeconomic level, financial markets react to changes in policy, and policy-makers react to the reaction of the market. Real economics is highly complex, recursive, and, crucially, involves the interaction of human psychology with changing market conditions. Economic life cannot be predicted or fully described by mathematics because it's just too rich and full of emotion, bias, love, aggression, and a near-infinity of other nuances of causation.

By trying to match the precision and language of science and claiming that all of ethics, psychology, and climate science can be captured in their models, many modern economists have become dangerously deluded in their ability to provide guidance on some of our toughest and most complex challenges. Canadian economist John Kenneth Galbraith captured this idea when he wrote, "Departments of economics are graduating a generation of *idiots savants,* brilliant at esoteric mathematics yet innocent of actual economic life."[124]

Nancy Cartwright, a philosopher of science I admired at the London School of Economics, had a favorite story she'd tell at the beginning of term. Imagine a piece of paper dropped from a high building in the middle of a city. Using only science and math, try to predict where that paper will end up. One can imagine large computers calculating the force of gravity battling with turbulent wind conditions, rotational dynamics, and who knows what else. Not easy to do, if possible at all. But if you knew that piece of paper was a twenty-dollar bill, you don't need much more than everyday common sense psychology to predict that it will end up in someone's pocket. The point Cartwright makes is this: sometimes common sense, which deals in broad generalities, can tell us more than complex theories that rely on precise analysis.

Anyone who watched Texas pucker up and dry out in the summer of 2011 (and 2012, and . . .) or who's had roses bloom in their garden during the Canadian winter knows something is wrong. When Pakistan has gargantuan floods for the second time in just a few years or Russia has a heat wave so intense that it temporarily halts grain shipments, our radar starts to flash a warning light. When Australia is hit year after year by floods, fire, and droughts of biblical proportions, our common sense tells us that something bad is happening. It doesn't take a great deal of imagination to extrapolate these first few decades of warming out a hundred years. We know in our gut it is not going to be pretty.

Yet economic models tell us not to worry.[125] The hubris of modern economics — its need to provide hard quantitative data and its inability to admit it doesn't know the answer to a problem — is leading us down a dark and dangerous path. By trying so hard to be like hard science, economics has become dangerously out of touch. It's like trying to predict the fall of that twenty-dollar bill without using psychology or common sense. In order to provide relevant advice on climate disruption, economics has to smarten up, admit what it doesn't know, stop being afraid of non-quantitative language, and start collaborating with other disciplines. What does that mean?

First off, get the science right. Climate science is honest about what it does and doesn't know. Uncertainties in the science are front and center. Scientists are allergic to hyperbolic statements, yet climate scientists are very, very worried. They are worried because, behind their conservative statements and admissions of uncertainty, many know full well there's a bad storm on the horizon. The uncertainty does not diminish their worry but increases it. Economics can adopt the best-in-class practices toward uncertainty and risk that science can provide.

Economics can admit to the severe limitations of its predictive power.

For example, with few notable exceptions, nobody predicted the massive financial tsunami that was the market meltdown of 2008. Princeton economist and Nobel laureate Paul Krugman wrote in 2009, "The central cause of the profession's failure was the desire for an all-encompassing, intellectually elegant approach that gave economists a chance to show off their mathematical prowess."[126] Simon Potter, director of economic research at the New York Federal Reserve, said, "Our profession hasn't reached the state where it can reliably forecast the economy. Our collective failure to do so in the financial crisis shows this." He goes on to say that we need to "ratchet down our forecasts."[127] In other words, forget that cost-benefit analysis that goes out a hundred years!

The messiness of human psychology—as a key component of economics—needs to be taken seriously using its own terms of reference. Sargent's work on rational expectation theory is a start. We can take seriously the idea that there is no such thing as value-free economics. It's just not true that human ethical choices can be determined by examining existing market activity. Human values must often be independently determined and imported into the model as an external variable. The discount rate, for example, is an ethical choice about how we treat future generations. If you want to know how people want to treat future generations, ask them!

The commitment to empirical observation as a basis for determining human values can be (and often is) taken to an absurd degree. It's one thing to say that you can tell *something* about people's values by observing what they decide to purchase or how they invest their money. It's something else entirely to say that that behavior fully determines their values.

If we determine ethical values exclusively by observing economic behavior, we commit a logical fallacy that strikes at the heart of what it is to be human. What we are becomes what is preordained; what we do defines what we will do. The logic is viciously circular. It completely ignores the possibility that we can choose our own future, can change our destiny, and can express our will to become something different than what we are.

We can also admit that some systemic risk is too much to bear. Some parts of our economy, mainly ecological, are simply too valuable to be put at risk. Functioning oceans are not up for grabs, nor is a stable weather system that we can count on to feed the human population. Some, like former Canadian prime minister Paul Martin, argue that we can value these things by putting a price on natural capital. Others, like British author George Monbiot, argue that putting a price on natural capital degrades the intrinsic beauty and value of nature. Still others, like climate expert Mike Hulme, say there is no price too high since the risk is systemic in nature. Reams of literature exist on the topic. It's called ecological economics. I'm not going to decide on a winner here, except to point out that it seems absurd to see economics as somehow separate from ecology. We are part of nature, not separate from it. The point underpinning all these positions is a simple one: traditional economics needs to get serious about new ways to value the systems that keep us alive.

Lastly, we can stop looking in the rearview mirror as a guide to what's ahead. History is useful and we can learn by looking to the past. However, this particular planetary experiment is one of a kind, and we might want to think twice about how much previous actions are a guide to the future. History is of little help in this coming storm. Nassim Taleb, author of *The Black Swan*, has written extensively about the human tendency to give far too much weight to the past, believing the old adage that history always repeats itself. It doesn't. His highly readable book provides many examples of pundits, analysts, and economists screwing up predictions of the future by looking too much to the past. Looking in the rearview mirror, we're headed straight into traffic.

So what's the alternative? How might these ideas result in an approach to economics that remains robust enough to provide real guidance? The danger in losing the precision of DICE is that we fall into a soft, mushy, feel-good kind of attitude—one that is honest about the complexity of the problem but unable to move through that complexity to real and justifiable action. Economics has the tools to provide real value in the debate about the proper response to the climate threat.

WAKING THE FROG: THINKING DIFFERENTLY — BUY INSURANCE

We know how to deal with risk. If someone told you the chances of your house burning down were tiny, you'd still buy fire insurance. It's worth paying a small, but not trivial, amount each year to cover your butt. Insurance is how we share catastrophic personal risk with the rest of the community. It's not only a good financial decision; we also get peace of mind. The way to think about climate policy is a bit like fire insurance. The cost we bear to de-carbon our economy faster than would happen under free market conditions is like buying insurance. We are insuring ourselves against catastrophic climate disruption, but there is one important difference. House fires are a risk; climate catastrophes are uncertainties.

Insurance companies can predict with great accuracy the probability of your house burning down. They use actuarial tables to do it—summaries of long historical trends about rates of fire, type and age of house, geographical location, etc. There are no actuarial tables for our warming world. Catastrophic climate disruption is a wild card. It's not possible to say with certainty how much we should pay in insurance, but we do have much better ways than DICE to estimate a reasonable amount of coverage.

We start by acknowledging, not ignoring, the uncertainties that cascade outward from the great unknown unknown of our one-of-a-kind planetary experiment. One suggestion is to let climate experts shape the probability distribution functions (PDFs) instead of letting climate novices (but economic experts) impose bell curves on the consensus best estimate. Having surveyed the relevant community of experts, we can map the range of best estimates and possible curves. Then, instead of aggregating those curves into a single PDF, we allow for a number of curves that best represent expert opinion for each causal step in the chain. On top of that, we might correct for well-known cognitive biases such as overconfidence.

This is just a fancy way of saying "let the range of uncertainty be defined by those who best understand it." The result fattens the tails of our probability functions, which means catastrophic events like extreme temperature rises are treated more seriously than previously.

We also admit that existing damage functions are arbitrary and only apply to the small changes in temperature we're the least worried about. Having fattened the tails of our probability function, we then take a good, hard look at what's under those fat tails. What does a 6°C (10.8°F) rise look like? A 10°C (18°F) rise? Of course, estimating the damages from these sorts of events is not something we can do with precision, but we can at least begin to talk about things like losing our ocean fisheries, turning our wheat fields into deserts, and so on. Those conversations force us to value ecological continuity and to understand that some resources (like arable land and bountiful oceans) cannot be easily substituted by technological advancements as prices rise.

We also acknowledge that setting the discount rate is an ethical choice. The proper discussion is not around what government bonds are paying and whether that rate is a better number to use than average stock market returns. We stop arguing about numbers and begin talking about what sort of risk we want to pass along to our grandkids.

We talk about our duty of care to future generations and use those discussions to set the discount rate. My bet is that when we look deep into our hearts, when we think about the sacrifices previous generations made for us, when we sit quietly in wonderment at the sort of delicate and breathtakingly beautiful ecological balance into which we and our civilization has evolved, that number will come out to zero. We treat future generations exactly as we treat ourselves.

Having framed the discussion in this way, we begin to understand what's really at stake. We begin to think about catastrophic climate disruption as something we simply cannot let happen. We begin to think about the planet like we think about our house. It's something we must insure against catastrophe. Of course, this kind of insurance is different. We can't ask our neighbors to pitch in when our planet burns (if such neighbors exist, they are far, far away in another galaxy). This changes the question from "How much *should* we spend on insurance?" to "How much *can* we spend?" While this doesn't give us a precise answer, we have begun to think in a completely new way about climate policy.

This is all very arm-waving I admit, but it can be made precise. The best effort to do so was the *Stern Review on the Economics of Climate Change*, commissioned by the U.K. Treasury under Prime Minister Tony Blair. Released in 2006, this report includes the possibility of catastrophic climate disruption and estimates its damage as a loss to global GDP of upward of twenty percent annually. Put more plainly, that is a collapse of our industrial economy. The costs of early and aggressive cuts in emissions are treated as a kind of insurance policy against this collapse.

The *Stern* report was the first mainstream economic analysis that took seriously the range of uncertainty in both the science and potential climate impacts. It also explicitly defended a very low discount rate on ethical grounds. Economist Nicholas Stern took a lot of heat for these two breakthroughs. Critics howled about the cost of the insurance

Stern recommended we buy, in the form of increased spending on low-carbon infrastructure. What were the critics howling about? The *Stern* report advocates spending somewhere between one and two percent of global GDP as insurance. That's peanuts.

Economists and pundits who would minimize what we spend on climate catastrophe insurance simply don't understand the nature of the threat. DICE ignores catastrophic climate disruption, arbitrarily limits calculations of damage to small, incremental changes in our economy and ecosystem, and discounts away to nearly nothing the concerns of future generations. Lomborg himself, along with mainstream talking heads on Fox News, has never bothered to understand the limitations of DICE.

After a day of debate with me, and several appearances together on Canadian television, Lomborg finally admitted (both to me and on air) that there may be something like "tipping points" in the climate system. It was like pulling teeth to get that much on the table, but I haven't seen him change his public position as a result, and he continues to provide comfort to those who would have us save our pennies rather than invest our dollars to lower the intensity of the climate storm. He hasn't demonstrated he really understands what those tipping points mean. Would you listen to someone who recommended you skip fire insurance?

SUGGESTED READING

Bjorn Lomborg, *Cool It: The Skeptical Environmentalist's Guide to Global Warming* (2007)

William Nordhaus, *The Climate Casino* (2013)

Jonathan Schlefer, *The Assumptions Economists Make* (2012)

Nicholas Stern, *The Global Deal: Climate Change and the Creation of a New Era of Progress and Prosperity* (2009)

CHAPTER 5
The Fossil Fuel Party

"The conventional wisdom protects the community in social thought and action. . . . But there are also grave drawbacks and even dangers in a system of thought which by its very nature and design avoids accommodation to circumstances until change is dramatically forced upon it."
— John Kenneth Galbraith, Canadian economist

The best parties are often followed by the worst hangovers, and the fossil fuel party has been a great one. Cheap, abundant energy powers modern civilization, bringing light, heat, and modern conveniences to billions of us. Nearly all of it comes from fossil fuels.[128] Coal, oil, and natural gas are almost single-handedly responsible for pulling humans out of the muck and into modern times. Without them, life would likely remain "nasty, brutish and short."[129] Plentiful hydrocarbons are why so many of us live so well.[130] Nobody wants that party to end, but end it must, and the sooner the better—the hangover we're already committed to is going to hurt!

Modern life sits atop a mountain of energy only kings could have had in times gone by. Our domestic appliances alone need the energy equivalent of four fit humans running hard all day, every day[131]—to say nothing of our cars, factories, planes, and office towers! Add it all up, and we have the equivalent of 150 virtual human energy servants at our fingertips.[132] Modern energy makes emperors of us all.

That infrastructure took generations and many trillions of dollars to build. It's indispensible, powering everything from jets and air conditioners to Internet servers and TVs. It built our skyscrapers, roads, and factories. It brings us fruit from Chile, wine from France, and iPods from China. Cheap and reliable energy underwrites modern civilization and penetrates every corner of our lives. It's a massive affair. Replacing it in a generation — which we must do if we are to avoid catastrophic climate disruption — is daunting.

Worse, the change we need flies in the face of accepted wisdom about infrastructure: slow and steady does it. The professional classes who manage, supply, and finance energy are paid to keep the existing machine running smoothly. They're not rewarded to risk charting a new course. The engineers who manage energy systems have to keep the lights on. The CEOs of energy companies are rewarded year after year for providing the same fuel. The bankers who deploy capital prefer to finance yesterday's technology, not tomorrow's. We respond by muddling along; we hope the party can last a bit longer, at least until someone somewhere sorts it all out. But that's wishful thinking.

There's an episode of *Magnum P.I.* — a show about an intrepid investigator I watched as a kid — that comes to mind. Magnum gets knocked off his jet ski by a careless boater. A circling shark adds to the drama. Magnum faces a danger too large for his trusty brawn or wit. He remembers a time his father told him to close his eyes and count to ten when things get too scary. The adult Magnum does just that. He counts to ten, opens his eyes, and lo and behold the shark is gone. In the real world we can't rely on luck and willful ignorance. The first reaction to danger is to understand what's needed to avoid it — even if it's a bit unnerving.

It's only natural we hesitate to call an end to the fossil fuel party, but a new and better party beckons. Low-carbon infrastructure can revitalize our moribund economy and ensure long-term economic stability. We just need our ambition to match the scale of the problem.

SIZE MATTERS ...

Oil tankers are some of the largest ships in the world. These industrial monsters stretch past half a kilometer (a quarter mile) and weigh up to half a billion kilos (1 billion pounds). Each of their several tanks is the size of a cathedral. Point them in the right direction, crank the massive 64,000-horsepower engines, and off they go. Their inertia is enormous. From a cruising speed of 29 km/h (18 mph), it takes a couple of kilometers and a quarter of an hour to bring these leviathans to a full stop. Their turning diameter is several kilometers wide. Not exactly light on their feet. Think of our energy system as that ship. Right now, it's headed straight toward climate disruption's stormy seas, so we need to change direction — and fast — but its sheer bulk means it's hard to change course quickly.

Globally, we spend about $5 trillion (between eight and ten percent of GDP) every year on energy, about ninety percent of it on fossil fuels. Almost 100 million barrels of oil *per day* are pumped from the ground and sent through a network of pipelines stretching hundreds of thousands of kilometers and onto thousands of tankers. In huge refineries it's "cracked" into the gasoline, diesel, and kerosene we burn in billions of engines and turbines around the world. Half of U.S. electrical power still comes from coal plants, whose appetite is satisfied by nothing less than blowing the tops off mountains to disgorge the coal within. Two coal plants a month go up in China. Huge new natural gas and oil reserves throughout North America are (literally) cracked open underneath our feet using new techniques to fracture the rock formations that hold it. There's so much natural gas coming from the ground now we hardly know what to do with the stuff. The amount of energy this global system delivers is staggering.

Say we want to reduce carbon emissions by eighty percent by 2030 to avoid the worst of the upcoming storm. How big a job is that?[133]

To replace eighty percent of North American oil we'd need around a thousand Olympic-sized swimming pools of biofuel each day.[134] That's impossible. There isn't enough biomass. Traditional sources of biofuel—from edible sources like corn or soy—are a non-starter. All the soybeans in North America turned into biodiesel won't make a dent in our diesel use—maybe five percent. We already use half of U.S. corn for modest amounts of ethanol. These sources help, but they can't replace our fuel.

We can also use non-food sources, like cellulosic biomass. Companies like Woodland Biofuels start with wood chips, agricultural waste, or almost any other sort of indigestible biomass. They rip it apart to form syngas—a mix of hydrogen and carbon monoxide. A series of chemical reactions, prodded by the right catalysts, produces fuel for your car. Cellulosic material might replace a third of North American oil.[135] We're going to need more efficient engines and lots of electric cars! Still, replacing just that much oil means a new plant—worth half a billion dollars—going up every few days for the next twenty years,[136] and we need infrastructure to collect and store all that biomass.

What about the electrical grid? I'm fond of pointing out there's enough solar in the American southwest or wind in Hudson Bay or geothermal beneath our feet to power industrial civilization many times over. It's true, but we need to build an awful lot of stuff to capture and transmit that energy. Just replacing U.S. coal—which accounts for half of U.S. electrical production—requires an industrial effort not seen since the Second World War, when factories were forced to turn from consumer goods to the war effort.[137]

Let's start with a couple of heroic assumptions. First, let's assume, even with lots of electric cars, that we cut our energy intensity in half.[138] That means when our economy doubles in size by 2030 (and by all account it had better!),[139] electrical use stays the same. Energy efficiency is crucial, but it can't get us there by itself. Second, let's assume all those electric cars balance their increased load with the storage we need to make intermittent renewables reliable, so they're a wash.[140]

How much wind, solar, and geothermal power would do the trick? If each of wind, solar, and geothermal power takes a third of existing coal plants out, we need something like 75,000 giant wind turbines, 25,000 ten-megawatt solar farms, and 150 massive geothermal plants[141] as well as transmission lines to connect them. That's an awful lot of rebar, cement, roads, wires, and rare earth metals, but it's not out of reach. We make seventy-three million cars and light trucks every year! These calculations are breezy but make a central point: the amount of stuff we need to produce is immense.

Other options can help. One is to replace coal with natural gas, which reduces carbon by half. It's already happening since natural gas is so cheap these days, but it's a stopgap. Natural gas can give us some breathing room but doesn't solve the problem.

Carbon sequestration or "carbon capture and storage" (CCS) captures carbon emissions from the stack of coal- or gas-fired plants and buries them deep underground in caverns or old oil fields.[142] The good news is you can keep using existing plants. The bad news is we need to build new infrastructure that compresses, transports, and stores the CO_2. To make a dent in emissions — say, a quarter of all stationary sources — we'd have to handle a volume that rivals global oil flows.[143] That costs about $300 billion a year, and we'd have to burn a third more coal to energize it. Plus we'd have to pay for the up-front cost of all that equipment—about half as much as the coal plants themselves. My own view is traditional CCS is a crock, more of an excuse for the industry to stay the current course than a real solution.

Even nuclear energy, hated by some but embraced by others as a workhorse of low-carbon energy, would be hard-pressed to pick up the slack. If global energy use increases as predicted, then a new nuclear plant needs to be built every couple of weeks from now until 2030 just for nuclear to maintain its current share of around fifteen percent of electrical production.[144] Some think that's a non-starter, if only because of a much-expanded nuclear waste problem.

I think we need to go next-generation nuclear—and fast. Breeder reactors can take existing nuclear waste and burn it multiple times, extracting an order of magnitude more energy than we got the first time around. Existing stockpiles of nuclear waste would last hundreds of years, and we'd be left with a manageable waste problem (see page 192). Still, we'd need to bring these new reactors online at a frenetic pace.

However you slice it, there's no free lunch. Any pathway — renewables, CCS, or nuclear — requires a huge industrial effort. Acknowledging the size of that effort is a good first step to delivering an appropriate response.

WRESTLING SIZE TO THE GROUND

For commentators like the University of Manitoba's professor emeritus Vaclav Smil, known as a hard realist on energy matters, these numbers indicate impossibility. His is a historical perspective: transitions from one form of energy to another—coal to oil or oil to natural gas—have taken many decades (often a century), even when the new technology is superior to the old. Energy's physical infrastructure has momentum, and it penetrates deeply into everything we do.

For analysts like physicist Cesare Marchetti, who was working back in the 1970s, patterns of human energy use seemed mathematically determined. Nothing we do can change the pace at which we move from one energy source to another. For Smil, we are historically determined: "the part of Marchetti's analysis that remains correct is the . . . extreme slowness of the substitutions, with about 100 years needed to [significantly change energy sources]."[145] Smil's conclusions are dire. He doesn't hold much hope our civilization will survive the century.

We don't need to go so placidly into that dark night. Our future is not determined. We can run civilization on clean, renewable energy, but we are wise to take heed of Smil's analysis. My interpretation of Smil is that, in the absence of historically unprecedented changes in market rules that prioritize low-carbon energy sources, we won't outrun previous energy transitions. Smil is correct to point out there is more momentum in energy systems than many optimistic commentators care to admit,[146] but to counter his warranted pessimism we must do one simple thing: decide to make climate security a top priority. It's that easy and that hard. But if we put climate security on an equal footing to military security—if we can just make that decision—our ambition will be matched by a modern industrial infrastructure and financial market of unprecedented might.

Global industrial capacity is much larger than it was during the Second World War, and it now includes the low-cost, high-volume production of China. Our scientific and engineering knowledge is far advanced, as is our ability to develop high-performance materials at scale. We've developed capacity for massively distributed energy systems, enabling the citizenry themselves to be engaged. The movement and degree of available capital is larger than at any time in human history, and the wizards of Wall Street might be encouraged to turn from self-inflating bubbles to matters of more general concern. Communications are global and instantaneous, and computing power has reached almost unthinkable levels. Let's just stop making so many toys and build things of more import.

However, to unleash that machine we must stare down a few hard truths. First, we have to cut the size of the job by using less energy. We need to cut it in half, oil even more, and that's not going to happen in a market economy unless we raise the price of energy. Second, without a historically unprecedented degree of market intervention, we'll fall prey to those slow historical rates of change. Third, we'll have to turn off existing plants before the end of their operating life and leave proven reserves of fuel in the ground. You can bet the owners of these so-called stranded assets won't be pleased. Fourth, a significant portion of our industrial infrastructure needs to be dedicated to the job. The supply chain of labor and materials we need is large enough to displace other production. We must collectively endorse the need to do it. We need broad public support. The siren song of denial remains a central character in our story.

INTERESTS MATTER ...

On a job this size, our biggest corporations must lead, not follow. Global energy companies have the engineering talent, financial heft, global reach, and industrial might to get the job done. But will they? Can we count on our largest global players to step up to the plate? In the absence of policy that forces them to, the answer is, unfortunately, no. It's difficult for captains of industry to even speak out on the topic.

In the summer of 2009, I had an opportunity to debate David Collyer, then head of the Canadian Association of Petroleum Producers

(CAPP), at an investment conference in Victoria, British Columbia. It was a debate I looked forward to. It's one thing to fire up a room full of people who agree with you. It's something else to make the case for a low-carbon economy to the head of an organization charged with developing high-carbon energy megaprojects. Here was a chance for a substantive discussion about the difficulties of de-carboning our economy with an expert capable of pushing back on my assumptions and arguments.

After some to and fro, I admitted fossil fuels would be difficult to replace. He agreed they were currently indispensible. I suggested we might start with a price on carbon. He said CAPP endorsed a small price. I noted that the urgency of the climate storm required more than a token price: we have to make difficult changes to our energy economy. The substantive discussion *begins* with how hard we're willing to try. Apparently not very, because at this point Collyer shrugged, "CAPP is agnostic on climate change."

I'd encountered the word agnostic before, in a debate with U.S. energy analyst Michael Economides on the Business News Network (BNN). After coming to the same potential point of engagement—it's hard but we have to try—he, too, shrugged and said, "I'm agnostic on climate change." At one level, declaring agnosticism is a good rhetorical trick to shut down debate while avoiding the embarrassment of denying climate disruption altogether. At another, it gives groups like CAPP tactical space: endorse a small and meaningless carbon price and deflect efforts away from a substantial one. The price on carbon becomes a straw man.[147]

But neither Economides's nor CAPP's agnosticism makes any sense. For CAPP, there's no reason to put *any* price on carbon unless climate disruption is real. When I pointed this out, Collyer disengaged entirely, and the debate fizzled to an end. Economides presents himself as a savvy twenty-first-century energy analyst, so he shouldn't be agnostic on climate disruption. It's like a pilot being agnostic as to whether the Earth is round: on short flights to the next city you can get away with it, but for anything longer it's fundamental. Economides is too smart to actually be agnostic on the biggest issue facing energy production and use in the twenty-first century.

There may be a more sinister motive at play. Agnosticism makes sense as cover to carry on with business as usual. It's a hedge against possible legal action. Corporate leaders, like everyone else, have a duty of care. They can't do things they know to be counter to the public good. Building new fossil-fuel infrastructure is exactly that. Plants built today have an operating life of many decades. It's almost certain we'll want to shut them down. Agnosticism provides a frontline defense against having to abandon those assets. A company can claim that at the time they made the decision to build it, they believed the jury was still out on climate change. Going on the record as being agnostic is not just a rhetorical device. It's a legal strategy, and a good one. From the point of view of a fossil fuel energy association, it's perfectly rational — if disingenuous.

Energy executives normally avoid debates on the climate and leave public declarations of agnosticism to lieutenants like Economides and Collyer. On the odd occasion they face questions on the climate, they tend to duck the issue. Rick George is a typical energy executive. An ex-CEO of Suncor and author of *Sun Rising*, he is a full-throated defender of Canadian tar sands development. Appearing in 2012 on TVOntario's *The Agenda*, he was asked several times about climate. After trying unsuccessfully to be subtle in his deflection, he flatly responded, "I don't like to talk about the whole climate change issue," and he changed the subject. Unhelpful, but not atypical.

Not all energy companies or executives are quite so slippery, but most are. Why the games? It will take time to get off fossil fuels, and energy companies are in the best position to take advantage of the transition: they can make money from the old and lead on the new. Why can't the broader energy sector publicly engage on climate? Rhetorical tricks and outright avoidance reveal a kind of corporate schizophrenia. Energy executives, even those who are well-meaning and cognizant of the impending storm, have no choice but to continue forcing their high-carbon product into the market. They have no choice but to work against real constraints on carbon. Executives cannot betray short-term profits for long-term benefits.

The robots in Isaac Asimov's famous book *I, Robot* (and in the subsequent movie starring Will Smith) are designed to never contravene their "prime directive": do not harm humans. Just like those robots, corporations have a prime directive: make money for shareholders. Their corporate documents normally describe how they will make that money: a coal company will mine coal, a car company will make cars, and a retailer will sell things to the public. Each takes advantage of its existing market position, assets, and skills. Corporations, and those who run them, cannot contravene this prime directive. Their shareholder agreements, governance structure, boards of directors—these legal structures are hardwired to ensure a company will do what it has set out to do.

It's not unreasonable that a coal company will fight to the bitter end and defend its right to mine and sell coal. Until market signals take away those profits, they *must* try to dig up every last piece of economically viable coal in the ground. That's what they are legally bound to do. Nor is it unreasonable to expect the Canadian Association of Petroleum Producers to do everything it can to delay a price on carbon, including tolerating a token price to deflect attention from the real thing. That's not a moral judgment but a statement of fact. They are just being rational actors in our economy.

That's why we have to use the power of the ballot box to change the rules of the game. The energy sector will not and cannot do this job of its own accord.

ATTITUDE MATTERS

"It is difficult to get a man to understand something, when his salary depends upon his not understanding it!"
— Upton Sinclair, U.S. author

The sad irony is that climate disruption will ruin things for everyone. Coal magnates, energy analysts, and shareholders of publicly traded corporations will all be worse off in the long term if we don't wrestle this problem to the ground. One might hope the long-term benefits of saving the ecosystem that keeps civilization humming might trump short-term motivations, however hard-wired they may be.

Corporate culture has its own inertia: the need to leave one's personal moral compass at home; the deep conservatism of large-scale capital; the willful blindness of those who feel insulated from risk; the habits of mind and conventional wisdom that build up during times of prosperity; and the self-selecting nature of those with their hands on the levers of power. These cultural forces limit us to small, incremental changes even while our most august institutions call for a more fundamental transformation.

John Kenneth Galbraith, one of the twentieth century's more erudite economists, coined the terms *culture of contentment* and *conventional wisdom* to describe how the received wisdom of the day becomes entrenched in ways that protect short-term affluence at the potential cost of long-term economic benefit. A simple way of capturing his insights might be: if you made a lot of money yesterday and it looks like you're going to do the same today, even tomorrow, why rock the boat?

Individual Conscience and Institutional Bias

Corporate social responsibility (CSR) and triple bottom lines (whereby financial, social, and environmental benefits go hand in hand) are attempts to weave moral considerations into corporate fabric. CSR is a good thing. It can focus a company on delivering the sort of good behavior customers expect: reduce pollution, treat employees with respect, act with integrity in jurisdictions without equivalent labor laws, lower energy use, and so on. CSR brings universally recognized moral frameworks into corporate boardrooms. The idea is good companies are rewarded with higher profits because customers prefer good companies to bad ones.

As long as customers reward corporations with higher profits, CSR will continue to change corporate behavior. But there's a limit. CSR is only effective to the extent that it increases profits. The rest is window dressing. It's good business to lower energy use and treat employees with respect. CSR is crucial in drawing attention to this fact. It may even be good business to put windmills in the field beside head office because it's worth the price of good public relations. But CSR by itself cannot change the rules of the game. A coal company will not become a wind company because it's the right thing to do; it will do so only when it's more profitable to make the switch. Corporations are amoral. Their priority will always be their prime directive of making (and increasing) profits.

People, on the other hand, have a conscience. We deliberate on both moral and practical grounds. We face difficult choices and balance many interests. That's what makes life so interesting. But corporations do not have a conscience: business leaders must often leave their conscience at home to do their jobs. Sometimes they can make small changes, such as lowering energy use to save money and carbon, but they cannot use their human values to fundamentally change what their companies do. If they tried, they'd be fired. This is the conundrum faced by many well-meaning corporate executives.

The CEO of an energy company might personally believe in climate disruption and worry for her children, but the short-term gains demanded by shareholders and enforced by her board of directors constrain her options. She's caught in a trap. If she leaves, someone less caring might take over, and she will lose what few cards she has to play. If she stays, she must play by the rules of the game and advocate for small, incremental change at best. It's not an easy position.

She likely can't even speak openly about her climate fear. When former chief economist of the Canadian Imperial Bank of Commerce Jeff Rubin decided to write about peak oil, he quit his job to do so. His were not opinions with which conservative banks wanted to be associated. His reaction is sanguine: "Looking back, I realize that McCaughey [CIBC's then-CEO] was only doing his job, which was to protect the bank's interests."[148] The ability of smart, insightful corporate leaders to have and express opinions that run counter to the interests of their employer is limited. That's why debates with heads of petroleum associations and big-time energy analysts are bound to be disappointing, no matter how bright those individuals may be. They're effectively muzzled.

Being part of something larger than oneself is a great thing. Our ability to form organizations is partly how we pulled ourselves from the muck. Organizations get things done that individuals cannot. We submit to its needs for good reason: "we surrender some of our autonomy in exchange for happiness, impact, and evolutionary survival. We surrender to the authority of the larger purpose, and in doing so, we become responsible *to* it."[149] But submission comes at a cost. Some of our brightest citizens are unable to contribute meaningfully to a broader public debate on the climate because their thoughts and concerns run counter to the prime directive of their organization. To fill the gap we use rhetorical gimmicks like "agnostic" to avoid that difficult place where morals and profits clash.

Willful Blindness: Other People's Risk

It's easy to turn to a blind eye to risks whose consequences are not ours to bear. That's why Wall Street's financial titans could ignore danger signs in financial markets prior to 2008. The financial crisis remains a potent metaphor for the coming climate storm. Many of our business elite remain willfully blind to the risks of climate disruption. Not unlike good old-fashioned denial, comfortable affluence keeps the risk from feeling up close and personal. Hurricane Sandy might have changed all that.

Given the implosion of the complex financial instruments Wall Street shoveled into global markets and the near-total meltdown they caused, one might be forgiven for thinking Wall Street financiers take risks, but they don't. Not really. When the U.S. government leveraged trillions to save the banks, it became clear the risk wasn't Wall Street's but ours. The most any leading financier stood to lose was a year-end bonus. Few CEOs' homes, livelihood, or credit at the nicest restaurants was at stake. Some of the smartest people in America were blind to the buildup of the most severe financial risk in living memory because they felt insulated from that risk.

Well before the crisis broke in the fall of 2008, it was clear something was wrong. Economist Nouriel Roubini sounded alarm bells in 2006 and a rank amateur set up a website called Implode-o-Meter to track sub-prime lenders as they went under, but the financial world marched on. Journalist and author Ron Suskind argues, "So many of the dynamics of the crisis . . . were exacerbated by the ego-addled dance of the CEOs . . . marking an era when the imperial chief executive often existed in a cloud city immune from accountability . . . the interests of chief executives were no longer woven with those of either shareholder or employee."[150] In her recent book *Willful Blindness: Why We Ignore the Obvious at Our Peril*, Margaret Heffernan argues that the major banks "did not see the risk because they did not want to. As long as everyone was making money, many CEOs either didn't see a reason to change, or lacked the courage to do so."[151]

Willful blindness is a well-established legal principle. Heffernan's analysis begins with the judge's instructions to the jury in the Enron trial, who said the accused "had knowledge of a fact if . . . [he] deliberately closed his eyes to what would otherwise have been obvious to him. Knowledge can be inferred if [he] deliberately blinded himself to the existence of a fact."[152] Over the years, juries have received this "ostrich" instruction countless times. Heffernan's is a fascinating analysis of why, and to what extent, people refuse to see what should be blindingly obvious — even something that will eventually bite them in the rear: "It's incredible, the arguments people come up with to perpetuate something that they know, deep down inside, is bad for them."[153]

At one level, it's simple: someone earning millions of dollars a year in salary and stock options, living in the modern equivalent of a castle, and possessing enough capital to ensure the financial stability of their progeny doesn't feel climate risk, even if they bother to think about it. Their world feels safe and insulated from food shortages, giant storms, rising oceans, and drought. Climate risk is not theirs to bear. To some extent, this is true. There appears to be a great distance between the lives of captains of industry and the effects of raging weather. That's why the flooding of New York City, including Wall Street, during the 2012 presidential race was so shocking. Hurricane Sandy crashed a party to which extreme weather wasn't supposed to be invited.

But with power comes more than feelings of security. Optimism, disconnectedness, an adherence to the status quo, and certainty, all are correlated with power. According to Frances Milliken of New York University's Stern School of Business, "We conceptualize power as a form of social distance and find that . . . power . . . was positively associated with the use of language that was more abstract . . . positive . . . and certain."[154] It's a heady brew that works to minimize threats like climate disruption and maximize opposition to change.

Optimism is a natural offshoot of power, partly because powerful people—from entrepreneurs who built companies to executives with lots of resources at their disposal—are accustomed to solving whatever problems come their way. Power and optimism are linked: power increases one's ability to solve problems, achieving a position of power affirms optimism, and striving for a position of power in the first place requires an optimistic outlook.

The more powerful you are, the more disconnected from others you tend to be. At one level, the disconnectedness is physical: the choice of limos over buses, first-class flights over economy, gated neighborhoods over apartment blocks, and all the other privileges that keep the powerful apart from the herd. Many do their best to avoid this isolation, but it's a broadly applied rule. With separation in daily life comes psychological distance. Life to the powerful appears smooth, not rough. Hostile opinions can be kept at bay. One rarely, if ever, must submit to the indignity of serving another.

This rarefied atmosphere affects judgment in obvious and subtle ways. Heffernan elaborates, "the psychological distance between themselves and others means . . . they have to think in far more abstract terms . . . and the combination of power, optimism and abstract thinking makes powerful people more certain."[155] The willful blindness of our business elite to elevated climate risk comes with the cheery hope of certainty.

A poster child for willful blindness is Michael Brown, who was the director of the Federal Emergency Management Agency (FEMA) during the aftermath of Hurricane Katrina in 2005. Brown appeared to be completely blind to the very risks for which he was responsible. He didn't seem to understand that people were hurting, even as he was being told it was the case on national television. His distance from the tragedy was both physical and psychological. His blindness to the situation was perfectly captured when he managed to joke in his now-infamous email: "Last hurrah was supposed to be Labor Day. I'm trapped now, please rescue me." None of that callous incompetence appeared to affect his confidence. Katrina is, of course, a stand-in for the climate storm ahead.

While there are many empathetic and competent leaders in positions of power, a malaise toward the climate threat has crept over the comfortably affluent. It doesn't feel close to home. The affluent have, by and large, demonstrated a commitment to the status quo over the radical change that's required. This is perfectly human; we all have a "preference for everything to stay the same. The gravitational pull of the status quo is strong—it feels easier and less risky, and it requires less mental and emotional energy to 'leave well enough alone.'"[156] But in playing to our short-term comfort we betray our long-term interests.

It's not just the business elite that can be willfully blind to climate risk. These arguments extend to many of us. The risks of climate disruption—food security, extreme weather events, drought, ecological viability, the Prairies reduced to a dust bowl—are distant, abstract. We're much more likely to be concerned about our job, our pension, our taxes, and our energy bills.

Public approval for action on climate is necessary in a democracy, but in a capitalist democracy it's crucial we have buy-in by our captains of industry. Without the influence, capital, lobbyists, business connections, and civic stature of our business elite fully aligned on carbon mitigation, we don't have all our oars in the water. Their influence on public debate can't be overstated.

Conventional Wisdom: Risk to the Status Quo

It's also easy to turn a blind eye to risk when it demands uncomfortable changes to the status quo—even when the long-term consequences *are* ours to bear. Economist John Kenneth Galbraith coined the term *conventional wisdom* to describe the cultural response to a threat to comfortable affluence. The threat here is not climate disruption itself but the required response. Many of us can't acknowledge the climate cliff ahead because it means an end to the fossil fuel party.

A radical rebuilding of our energy economy is understandably disturbing to those who benefit from its comforts, which includes all of us! For Galbraith, ideas become acceptable to broader culture on the basis that they defend existing comfort: "Numerous factors contribute to the acceptability of ideas . . . we associate truth with convenience—with what most closely accords with self-interest and personal well-being or promises best to avoid awkward effort or unwelcome dislocation of life."[157] Climate disruption's threat is long term. Having to turn off coal plants is more immediate. We prefer ideas that defend against the latter, not the former.

Galbraith argued conventional wisdom can subvert self-interest and was worried about the "grave drawbacks and even dangers in a system of thought, which by its very nature and design avoids accommodation to circumstances until change is dramatically forced upon it."[158] He didn't mince words: "the fortunate and favored, it is more than evident, do not contemplate and respond to their own longer-run well-being. Rather, they respond, and powerfully, to immediate comfort and contentment."[159] He cites the pushback against Roosevelt's New Deal over its "conflict with free market principles, its impairment of essential economic motivation . . . and its seeming subversion of sound money and public finance."[160] In retrospect, it's understood that "the Roosevelt revolution saved the traditional capitalist economic system and the well-being of those whom capitalism most favored."[161]

The New Deal might be old news, but former Federal Reserve chairman Alan Greenspan's refusal to regulate financial markets right up until they fell off a cliff reverberates today. In the face of counterevidence like the dissolution of Long-Term Capital Management back in 1994, the lure of defending a comfortable status quo that was addicted to unregulated production of complex derivatives proved too much to resist. Too many people were making too much money. To the end, while markets crashed down around him, the most he could admit was that he had found "a flaw."[162]

The conventional wisdom of the time embraced free markets, and lots of smart people were blind to two important truths: markets aren't truly free, and there are huge downsides to a free-for-all. The long-term self-interest of the financial system was betrayed by a conventional wisdom that sought to keep the status quo. Galbraith's writing sounds prescient: "What is not accepted . . . is the powerful tendency of the economic system to turn damagingly . . . ruthlessly inward on itself."[163] As go the financial markets, so goes the climate, but this time there's no Federal Reserve to give us another chance.

Conventional wisdom builds on itself. Business schools churn out new managers that are a lot like the old ones. Captains of industry are a self-selecting group. The closer you get to the levers of power, the more likely it is you'll need to demonstrate a commitment to the goals of that machine. You don't get to be an investment banker without reciting a few free-market mantras along the way or an oil executive without at least tacitly agreeing to stay mum on climate. The new boss really is the same as the old boss.

But events catch up with spin. Reality always bats last: "The enemy of the conventional wisdom is not ideas but the march of events."[164] Hurricane Sandy was a reality check that washed against the conventional wisdom that had lulled the presidential election into complacency on climate.

There are lots of people who are deeply concerned about climate and buck against the indifference of their peers. I've had the pleasure of meeting many of them. Canada's Greg Kiessling founded Bullfrog Power to teach the market about clean energy. Silicon Valley's Vinod Khosla makes big, bold bets on clean technology. People such as Bill Gates, Richard Branson, and Elon Musk are clearly out to make the world a better place. All these (and many more in many walks of life) are wonderful exceptions, but it's a numbers game. For every Kiessling, there are a dozen Koch brothers. For every Khosla, there are lots of Economides. For every CEO brave enough to speak out on the issue, there are dozens who stay quiet — and silence is the language of inertia. These exceptions just confirm the rule.

Until we feel the risks of climate disruption are shared by us all, we will likely remain seduced by a conventional wisdom that defers uncomfortable action to another day. Without a concerted effort by rabble-rousers from all walks of life, the conventional wisdom will morph from outright denial into denying the degree of danger and straight through to "it's too late to bother trying." Unless we push hard, there may never come a time when concerted action is the wisdom of the day. It's not hard to see coming. It's happening now.

Shale Gas: Progress Trap?

Imagine a fish happily following a tasty trail of aquatic treats along a tidal flat. Each flick of its tail lands it a reward. But as the tide slowly recedes, it finds itself trapped in an ever-shrinking puddle. It's unable to get back to the ocean. Before the high tide returns, the sun evaporates what little water remains. The fish dies. That poor fish fell into a progress trap. A progress trap is a series of moves where each seems to make sense on its own, but together they take you to a nasty place. Each step feels like progress, but the path is a regression. Every decision is locally rational but adds up to error.

Fossil fuels are a progress trap. Each step forward seems to be in our best interest: we build another gas plant, drill another oil well, or burn another ton of coal—after all, that's how life got so good in the first place—but enough steps in that direction take us to economic ruin. What we really need is a new path, not just small changes to the status quo. Shale gas—cheap, plentiful natural gas released by fracking rock—is a bit of a wild card: part progress trap, part new pathway.

Recall that a modern, dynamic description of the economy emphasizes its creative potential while acknowledging its path dependency. Each step affects what can happen next, for better or worse. Some steps open new possibilities, others close them off. An economy with lots of operating coal mines, for example, is more likely to build more coal plants since the mines already exist. The mines appear to be an asset but are actually a long-term liability because they deter investment in alternatives. Existing highways make it easier to encourage the production of more cars rather than to make the capital investment in a train track. And so on. What's already happened defines what might happen next.

Natural gas is sometimes called a bridge fuel to a low-carbon future because it has half the carbon content of coal. One response to climate disruption is to crank up the dial on natural gas. Certainly it's a move heartily endorsed by the fossil fuel industry. A typical headline reads, "Exxon executive urges Europe to find shale gas."[165] The ability to fracture rock and release previously unrecoverable natural gas reserves has brought prices in North America to record lows, so it also makes short-term economic sense. America's recent shift from coal to natural gas dropped U.S. emissions for the first time in history. While there are concerns around the fracking techniques that are unleashing that torrent of natural gas, like water impacts and gas leaks into the atmosphere, with strong regulation it seems like a win-win—lower costs and lower carbon.

But if we replace coal with natural gas plants instead of making investments in zero-carbon energy, we're in a progress trap. We need to drop emissions in the electricity sector by more than half. Natural gas investment, by itself, still locks us into a high-carbon future. If natural gas is a bridge fuel, it has to be a bridge to somewhere. That place is a low-carbon grid dominated by clean, renewable energy, and natural gas has a more important role to play than just acting as a lower-carbon coal.

Natural gas enables renewable energy. Modern natural gas plants can crank their output up and down really fast. That's useful—they can act as a stabilizing force for an onslaught of highly variable renewables like wind and solar. We can leverage natural gas to get us up and over a hump to someplace really interesting. Unlike that poor fish, we'd be headed for the open ocean. But if all we do is use it to replace coal, we're just buying a bit of time. Like the fish, we're tempted to continue in the wrong direction.

By thinking in terms of incremental change, we sometimes rule out completely different pathways that could take us to a much better place. Let's imagine what the final result might like and just go for it!

Waking the Frog: Taking Our Energy Moon Shot

We still have time for an energy "moon shot," a publicly directed new low-carbon energy mission that unlocks the engineering, industrial, and financial might of the global market economy. It will rebuild our energy systems by giant leaps, not incremental changes. The mission will have a clear end point: a low- to zero-carbon future that delivers on an aggressive temperature target. It sharply accelerates all low-carbon pathways: efficiency, next-generation nuclear, enhanced geothermal, carbon capture and storage, and both centralized and distributed renewables. The energy moon shot unlocks the might of the private sector but sets the pace and scale of change. It also comes at a time when the moribund economy needs a shot in the arm. This can be our Apollo moment!

It's a huge job, much more ambitious in scope than the original moon shot. Market interference bold enough to upset the historical applecart will align corporate self-interest and climate well-being to bring unprecedented industrial might to bear on this singularly important task. We can unlock the massive pools of capital that sit in pension funds, and we will support energy and business executives willing to stick their neck out to take a strong public stand on climate disruption—even if it goes against the conventional wisdom of their peers.

The energy moon shot is a multi-stakeholder effort. Everyone rows in the same direction. Science sets the goal. Governments provide the right policy support. Industry's job is to get us there, not to obfuscate the science or lobby against the policy. Industrialists, financiers, and captains of industry contribute to the debate and start by saying, "We get it. We're on it." We have hard choices to make, and we need them to put their collective wisdom forward in good faith. What we don't need are declarations of climate agnosticism. Such rhetorical gimmicks are both churlish and childish and abdicate any claim to civic leadership.

The only fight worth having is about the *details* of our new moon shot—not whether the moon exists, is made of cheese, or is too far away. We can and should argue about natural gas versus renewables, how big a role nuclear or carbon sequestration will play, whether a carbon tax is better than cap and trade, and whether or not either should be revenue neutral.

The moon shot is a local, national, and international priority on a par with personal and military security. Look no further than our military for leadership: there is an awful lot of capital, know-how, capability, and technology in standing military forces largely devoted to keeping oil supplies running. The Middle East is, at least in part, a hot spot because it sits on oil. Get away from oil and free those forces for peacetime contribution to the energy moon shot. Militaries around the world have a mandate to secure our collective future and are legitimately biting at the bit to contribute to a climate solution. They know what's at stake. Our military industrial complex can be a strong partner.

Sure it's a long shot, but that won't stop us. U.S. president John F. Kennedy knew how far away the moon was when he committed America to send a man there. It was an almost unimaginable distance, but we still did it. We can do this too, if we decide it's a priority.

How we do it is the subject of the next chapter.

SUGGESTED READING

Lester R. Brown, *Plan B 4.0: Mobilizing to Save Civilization* (2009)

Tom Butler and George Wuerthner, *Energy: Overdevelopment and the Delusion of Endless Growth* (2012)

John Kenneth Galbraith, *The Affluent Society* (1998) and *The Culture of Contentment* (1993)

Vaclav Smil, *Energy Transitions: History, Requirements, Prospects* (2010)

Daniel Yergin, *The Quest: Energy, Security, and the Remaking of the Modern World* (2011) and *The Prize: The Epic Quest for Oil, Money and Power* (2008)

CHAPTER 6
Waking the Frog and Turning Down the Heat

"Nature, to be commanded, must be obeyed."
— Francis Bacon, British philosopher

Our frog isn't moving much, but it's not yet paralyzed. The pot is hot and will get hotter, but it's not yet boiling.[166] We can still act, but time is running short. There are two ways this story can end.

In one, we continue with business as usual. We remain in blissful ignorance, enjoy the wealth we have created, and ignore for as long as possible the buildup of carbon and our increasingly unfriendly weather. We continue to order takeout and watch crazy storms and droughts on the Weather Channel. We feel badly for the people paddling in little boats down what was once their street or sitting forlornly in a dusty field, but we push the concern to the back of our mind. After all, we've got our own worries—those other problems seem so far away. Maybe we start to notice something's wrong by a sudden spike in the price of the pizza that arrives at our front door, caused by simultaneous droughts in the U.S., Australia, and Russia that dry up global grain markets. By then it's too late.

At that point we (and our kids) can only hunker down as best we can while the grandmother of storms and granddaddy of droughts hit us again and again, without end. Our economy will be stretched to the limit as we try to adapt: we build sea walls, new bridges, move towns, and, most importantly, figure out new ways and places to grow food. This story is a dystopia. Our way of life doesn't survive the century. Not the best choice—we'll all agree in retrospect!—but it's the one we're *already making* by not consciously, and determinedly, choosing a different future. This story is our default because it's the decisions we make now that set our course for many decades, if not centuries, to come. And the party's already started. Hurricanes Katrina and Sandy, the deep droughts in the United States, Pakistan, and Russia, the disappearing Maldives— these are but opening acts for the main event.

This bleak dystopia is not just my opinion. It's the view of some of our largest, most trusted, and conservative institutions: the Pentagon, PricewaterhouseCoopers (PwC), NASA, the International Energy Agency, and a host of others. These aren't groups prone to environmental hyperbole. If, after all you've read here, you prefer the "don't worry be happy" message of the *Wall Street Journal* or Fox News (both part of the Rupert Murdoch empire), I doubt there's much anyone can do to convince you otherwise. But I urge you, dear reader, to take the time to read the reports I've identified here. Behind the somewhat dry, technical language, alarm bells are ringing.

On the other hand, we might slow the warming and squeak our way through to a slightly warmer place, but at least it will be one that continues to support civilization. This story involves *all of us* as its authors. How might that story go . . . ?

It would start today. While pundits on Fox keep up the charade that the science isn't settled, public opinion quietly shifted. It started with Hurricane Sandy hitting Wall Street and was followed by President Obama's appeal to common sense in his 2013 State of the Union address. Then pieces started fitting together. Evidence of a burning Australia, melting Arctic, dried-out Canadian Prairies, flooded Calgary and Toronto, and decimated American cattle herds break through our collective cognitive defenses. Deniers begin to look downright silly. Church leaders start speaking of a new responsibility to God's green Earth. Children of corporate leaders bring uncomfortable questions to the family dinner table. Unwilling to be a villain in their kid's eyes, that discussion quietly sneaks into boardrooms around the world.

But it is the Great Global Drought of 2016 that changes everything, when grain prices skyrocket around the world. Food prices escalate, and our farming sector collapses. Even in North America, many grocery shelves are empty, rice imports having been stopped as governments around the world scramble to feed their own people. Food riots lead to very different results at the ballot box. Business leaders are forced to explain to a frightened and angry public what they are doing to solve the problem. Those with a track record on climate action double their market share overnight. Those who have exacerbated the problem see their revenues plummet. People vote for action at the ballot box and with their wallets. No longer do vested interests fight strong climate law.

Now, imagine it's 2020 . . . Packaged goods have a carbon footprint labeled beside the bar code. Supply chains have responded "Walmart style" to the demands of large retailers,[167] bringing a lowered carbon footprint to their customers' loading dock. Pension funds have acknowledged they can't meet future obligations in a hotter world and have begun pouring trillions of dollars of long-term money into low-carbon infrastructure. The moribund economy of the developed world bounces back as the massive stimulus effect kicks in.

There are economic casualties . . . Share prices of fossil fuel companies, historically based on proven reserves, plummet as investors realize those profits have to be left in the ground. Those prepared for the shift, and willing to move quickly, evolve into clean-energy companies and regain market share. Their employees take advantage of pension-fund-backed retraining programs to fill the huge gap of skilled trades. Others see their coal plants shuttered and revenues dry up. The lawsuits brought by angry CEOs are thrown out of court on the basis that they should have been aware of the carbon risk a decade earlier.

By 2040 we'll have averted the worst of it and squeaked through by limiting warming to 3°C (5.4°F). A new and sustainable relationship between our energy systems and the planet opens the possibility of another thousand years of civilization. Where that takes us, nobody knows!

This imagined scenario is not quite a utopia, especially if you happen to be a coal miner or shareholder of fossil fuel stocks. But this story ends with a chance at a few more chapters. How we get there from here remains to be seen, but we can start with some basics.

WAKING UP

We know why we're paralyzed. From the siren song of denial to economic models that discount the future and simplify the science and a free-market ideology threatened by climate disruption to the conservative tendencies of capital and our economic elite, our paralysis is more than a bunch of bad guys rigging the game—the rules of the game itself are predisposed to ignore climate disruption. Now that we can see how those rules operate, we can work to change them and tilt the odds in favor of action.

We start by silencing the siren song of denial. The truth is on our side, but our shared cognitive biases mean rational dialogue is not enough to wake the frog. We must find new ways of speaking and leverage those same cognitive biases to work in favor of action. Climate action is associated with a new era of renewed prosperity. We can react to climate disruption with a sense of possibility and purpose. We can visualize a brave new world of clean energy abundance and sustainable economic growth instead of deprivation and despair.

That conversation could spread like wildfire, sidestepping existing media silos that have proven so unable to communicate honestly with the public. Google Earth and World Wildlife Fund partnered to create the "vote Earth" initiative, which distributes an electronic ballot box to computers and mobile phones. More than three billion mobile phones can now be leveraged into a single voice for action—no mainstream media required! It's of limited pragmatic force, but, used effectively, it can generate moral suasion for targeted political leaders. If the secretary general of the U.N., U.S. president, or even a single recalcitrant senator, congressman, or member of Parliament gets hit with a few billion unique "act now!" electronic triggers, they'll feel a moral mandate to act. Citizens would thereby get renewed strength knowing they're not alone. It's easy to sink into lethargy when our collective punditry keeps saying action is impossible. Billions of electronic voices, singing in harmony, can lift us above that imposed pessimism.

Over time, climate denial could become as uncool as racism. The cultural shift would be accompanied by new social norms. Reminders of our awakening will be all around us—in TV shows, advertising, toys, even the status symbols we use to show off. We can co-opt the natural human desire to conform to new models of success. Today's status symbols of high-carbon living would be turned into emblems of ridicule. Driving a Hummer down a city street will be like wearing fur to an animal rights party. There'll always remain some who are unconvinced, but they'll get dragged kicking and screaming into the twenty-first century.

We can push aside the simplistic myth of the free market that remains a hurdle for so many on the political right. There's no contradiction in simultaneously recognizing the power of the market and the need to tame it to our needs. It's a bit like fossil fuels: without control they're destructive, but harnessed they provide useful work. We can stop viewing the market through the lens of an eighteenth-century idea of equilibrium and use twenty-first-century thinking. The global economy is a highly complex, non-linear dynamic system that constantly reinvents itself. The way it evolves to open new possibilities can't be predicted with precision, but it can be set on a general course. We can engineer market rules that move us to a low-carbon economy. The most effective tool in our arsenal is a global market signal that affects all transactions at every level of exchange: only a price on carbon can simultaneously harness the market and unleash its creative potential. The political right and left can come together on market-based solutions.

With a carbon price in place, our entrepreneurs, financiers, and engineers can get to work. Without it, the rip 'n' burn attitude to resource extraction will continue unabated. Imagine if we motivated all that brashness and brainpower on Wall Street to solve the carbon problem: instead of building sophisticated models to arbitrage millisecond moves in asset prices, we'd constantly identify the most efficient path for capital to mitigate carbon across sectors, geographies, and technologies. That's the market working *for* the common good, not against it.

We can build more sophisticated (and honest) economic models to guide policy decisions. They would include a collaborative, multi-disciplinary approach to climate science, risk management, and uncertainty. The discount rate is set to zero to emphasize what we really want: long-term economic stability and prosperity that includes our grandkids. The cost of climate response can be seen as insurance against long-term economic and social turmoil, not friction in an otherwise flawless economic machine. We could use the insight of economists, but their discipline would become more honest about limitations to quantitative analysis, more collaborative, and better able to accept input on moral grounds.

We could be honest about the scale and pace of change needed to avert the climate crisis. It's a huge job; pretending otherwise lulls us back to sleep. Incremental change can do more harm than good by leading us down a blind alley. Small changes sucker us into thinking we're doing what's needed when we're not coming close. It's a bit like putting ten percent ethanol blend into your SUV or changing a few lightbulbs: we feel good enough to turn away from the harder job that remains.

Better to go for the moon shot right out of the gate — rebuilding our entire energy system in a couple of decades. A lot of capital has to move quickly. The historical conservatism of debt capital can't dictate the future pace of investment. Existing fossil fuel plants will need to be shut down before the end of their operating life. Fossil fuel resources will have to say in the ground, untapped, with their profits unrealized. We'll stare down those fossil fuel giants: they either commit their resources to that moon shot or risk irrelevance.

The new normal in business schools and country clubs will be to admit the possibility of limits. Rather than the current rip 'n' burn business ethic, our business leaders will understand the real long-term economic benefits of sustainability, and they'll be able to act on those beliefs because we'll change the rules, ideally with a price on carbon. The balance sheets of energy companies will reflect their long-term prospects in a carbon-constrained world. Over time, we'll collectively demonstrate ever more clever ways to collaborate with, rather than overcome, nature. With some humility, we'll recognize the limits to how human ingenuity and natural capital can interact.

Finally, our elite will step up to the plate en masse and shake themselves out of their collective complacency. The climate problem is one that we all face. Their attitude will no longer be one of ambivalence, or agnosticism, but one that says, "I get it. We're on it." There are no gated neighborhoods in the global commons. I cannot imagine what sort of world the Koch brothers will inherit if they have their way, or what fun it would be playing king of the hill there. What's the point of being a captain of industry and providing a nest egg for your progeny if the world in which they wake up is hostile, barren, horrific?

We know why we're paralyzed. Acting on that knowledge wakes up the frog. How then, do we turn down the heat?

Turning Down the Heat

Greenhouse gases—particularly carbon dioxide (CO_2)—are heating the planet. Just like glass on a greenhouse, they trap heat from the sun. It's getting uncomfortably hot, and the temperature will soon reach dangerous levels. We know we can lower the heat by dialing down our greenhouse gas emissions, and we know how to turn the dial.

Like any journey, we begin by picking a place to go and work back from there. We start by setting a temperature target as an upper limit on warming. Because the science is expressed in probabilities, we refine the target as the odds of going past that upper limit. Climate science translates our target to a maximum level of atmospheric carbon (in parts per million, or ppm) using an estimate for climate sensitivity (a measure of how much the planet heats up as carbon levels increase). Climate sensitivity is normally given as the final, steady stated temperature increase that results from a doubling of pre-industrial carbon levels. With a final target level for atmospheric carbon, we work out a range of allowable emissions pathways (total annual global emissions).

A simple example will help. Let's call the Rand scenario a fifty-fifty chance—like tossing a coin—of shooting past 3°C (5.4°F). This is a very dangerous level of warming and represents one of the gloomiest scenarios we might consider. Given a climate sensitivity of 3°C (5.4°F),[168] we calculate the maximum carbon level associated with a fifty-fifty chance [169] of going past our target: 560 ppm CO_2e,[170] or a doubling of pre-industrial levels (in this example, the climate sensitivity is equal to the temperature target to keep it simple). Since we're at 430 ppm now, we're allowed future emissions that add a total of 130 ppm to the atmosphere, or around 2,600 gigatonnes (about 2.8 trillion tons).[171] Current emissions add 2.5 ppm CO_2e (50 Gt/55 billion tons) annually and are rising around three percent per year.[172] These are *net* emissions, since not all our emissions stay in the atmosphere. Between a third and a half are reabsorbed by the oceans, for example.[173]

There's a range of emissions pathways allowed under the gloomy Rand scenario. The sooner our emissions peak and start going down, the more slowly we can reduce emissions. The later they peak, the more drastic those cuts have to be. The timing of our peak emissions is absolutely critical—such are the wonders of compound growth. At one extreme, we continue with business as usual and hit our limit in just thirty years, at which point we'll have to drop them to zero overnight—which is obviously impossible. On the other hand, if we start dropping emissions tomorrow (making the peak today), even a small annual decline of less than one percent keeps us within the limits of the Rand scenario until well past 2050.[174]

The Rand scenario clearly shows early cuts matter more than later cuts. It's important we hit our emissions peak soon. But emissions today are not just rising, they're rising at an increasing rate—we're accelerating in the wrong direction! We have little time left to turn that trend around, but the good news is if we do peak quickly, there are lots of emissions pathways that give our economy a chance at a soft landing.

The Rand scenario might sound reasonable—we limit warming to 3°C (5.4°F) or so—but it's very dangerous. The uncertainty in the science, expressed as the probability of avoiding the temperature target, hides some scary stuff. A fifty-fifty chance of avoiding 3°C (5.4°F) means there's still a reasonable chance we hit 4°C (7.2°F), 5°C (9°F), or more. What are those odds? It's anyone's guess and depends on the shape of the probability curve (as we saw on page 126), but an expert best guess puts the odds of going past 4.5°C (8.1°F) at twenty-three percent.[175] That's why most proposals use a small probability of going past the target temperature, which implies a lower carbon ceiling. In the Rand scenario, if we target less than a five percent chance of exceeding 3°C (5.4°F)—a *real limit* of 3°C (5.4°F)—our carbon limit is nearer 400 ppm.[176] We're already past that.

We're late to the game and have let carbon levels rise too high already. We can't just muddle along and hope for the best, so the target temperature takes priority. We start by limiting risk to a manageable level then commit to policy choices that get us there. It's no longer appropriate to ask, "How much can we afford to spend on climate mitigation?" The question of relevance is now "How much do we have to spend to get where we need to be?" First pick the target then commit the necessary policy and capital.

The Rand scenario is a vast oversimplification, but the lessons apply to *any* effort to turn down the heat: our emissions must peak soon, any target that limits risk to acceptable levels is going to be very aggressive, and a target temperature to limit risk takes priority over other considerations. There are lots of comprehensive proposals that include complexity that I've skipped:[177] the IPCC, the International Energy Agency (IEA), the *Stern Review*, the European Commission, and so on.

All scenarios agree on the importance of an early peak in emissions. The IEA has gone so far as to quantify the cost of delay, reporting in their *World Energy Outlook 2011*: "Delaying action is a false economy: for every $1 of investment avoided in the power sector before 2020 an additional $4.30 would need to be spent after 2020 to compensate for the increased emissions."[178] This view is echoed by a number of studies on the costs of responding: "We find that political choices that delay mitigation have the largest effect on the cost–risk distribution."[179]

Most serious scenarios set a target temperature of 2°C (3.6°F) as our line in the sand because there's a danger that, past that temperature difference, positive feedbacks take us to much higher temperatures no matter what we do. The carbon level most often associated with warming of 2°C (3.6°F) is 450 ppm, which makes it seem as if we might make it, but we likely cannot. A carbon level of 450 ppm is a three-quarters chance we exceed 2°C (3.6°F).[180] The prognosis for 2°C (3.6°F) doesn't look good: "In the absence of any serious mitigation efforts . . . the likelihood of limiting warming to less than 2°C (3.6°F) is essentially zero."[181] The 2°C (3.6°F) goal represents political expediency more than real science.

To have a snowball's chance of limiting carbon to 450 ppm, we have to leave between two-thirds and four-fifths of all proven fossil fuel reserves in the ground.[182] Proven reserves are the stuff energy companies intend to extract and sell. They sit on corporate balance sheets and set stock prices. It's what the market tells us will be burned, but we can't burn them. The math is simple: we can emit maybe 600 Gt (600 billion tons) more of carbon dioxide before we hit 450 ppm. Burning proven reserves will emit almost 3,000 Gt (3 trillion tons). The only way out might be massive (and immediate) deployment of carbon capture and storage.

My view is harder and mirrors the organization 350.org: carbon levels must be brought *down* to 350 ppm to avoid positive feedbacks. The findings come from top climate scientist Jim Hansen's work.[183] Intuitively, it's simple: northern permafrost and ice cover have already started to melt. Both cause more warming. Once that process starts, why would it stop? We need to go backward and reduce both emissions and the absolute level of atmospheric carbon. That means sucking it out of the air, or geo-engineering, to give us some breathing room. Not great news, I'm afraid. In any case, the target has to be aggressive.

So how do we turn down the carbon dial?

A Fourth Industrial Revolution

The first three industrial revolutions—coal-powered steam engines, oil-powered mass transport, and microchip-powered information systems—came from technology unlocked by science. Thermodynamics (and a supply of coal or oil) unlocked the first two, and quantum mechanics (and a supply of semiconductors) enabled the third. We got the steam engine, the car, the Internet. The pattern is similar: science enables technology, which gives us control over nature, leading to the promise of a better life and stronger economy. The market emerges somewhat spontaneously[184] because the promise of a better life is immediate and gratifying.

German chancellor Angela Merkel was one of the first to call our transition to a low-carbon economy the fourth industrial revolution. But this time the pattern is different. Instead of promising a better life, science has identified the threat of a diminished one. Instead of controlling nature, we need to stop influencing it. Incumbent technologies seem fine over the short term, so the gratification from new, clean alternatives is delayed.

Our challenge now is to create a market to unlock the low-carbon economic revolution. We know how to do that: the centerpiece is a price on carbon.

A Price on Carbon

There's no more powerful tool in our policy arsenal to unlock a clean-energy revolution than a price on carbon. All other options — feed-in tariffs, regulation, green bonds, subsidies, and penalties — play second fiddle. Pricing carbon is fair, justified, effective, efficient, and politically neutral.

A carbon price is fair. It passes the most laissez-faire of economic tests: no one gets a free ride. Everyone agrees we can't dump garbage for free in a field. It's only fair we pay, in this case by using the city dump. Similarly, we can't dump carbon into the atmosphere for free. Just because you can't see it or smell it doesn't mean it's not pollution. The cost of carbon is currently externalized — the harm it causes is passed onto the public at no cost to the polluter. A carbon price internalizes this cost so a product's manufacturing cost reflects its real cost. A principle of climate fairness is the polluter pays.

A carbon price is justified. It reduces risky behavior that affects us all. Nasty stuff happens when systemic risk is ignored, and it's the public who pick up the tab. Systemic risk builds throughout a structure until — wham! — something goes wrong and the whole thing shakes. If I risk my mortgage in Vegas, the effects are limited to me. There is no systemic risk. When financial firms ignored the risk of complex debt instruments, it spread all over the interconnected financial system.[185] Suddenly, the system shook: liquidity dried up, the economy shrank, retirement plans were shot, etc. Wall Street didn't pay the price; we all did with a government bailout and shrinking economy.

Climate risk is no different. It builds slowly, throughout the global ecosystem and economy. When the climate shakes — with increased extreme weather events like Hurricane Sandy, increased levels of droughts and wildfires, rising oceans, etc. — it's not the big emitters who pay, we all do.[186] When carbon isn't priced, there's no incentive for emitters to limit the risk we all share. A principle of climate justice is shared risk must be priced appropriately. We can add another: money raised from shared risk pays for shared damage. Justice and fairness are related.

A carbon price is effective. Four principles of a dynamic economy are the market is unmatched in creative potential, we have overarching goals toward which we should aim, we cannot effectively pick technological winners and losers independent of the market, and early actions matter because the evolving economy is path dependent (see page 108). A price on carbon is the only policy option that addresses all four by simultaneously affecting all market transactions: the daily retail choices of consumers, the long-term investment decision of pension funds, the sorts of technologies venture firms choose to back, and so on.

A carbon price tames the market and harnesses its creative forces to optimize for the constraint of lowering carbon. It sets new rules, and the private sector acts on them any way it chooses. A carbon price doesn't pick winners and losers, but it defines how they're chosen. Of course low-carbon infrastructure is set to be the ultimate winner, that's the direction we've chosen, but the way we get there is unknown. The creative potential of trillions of market transactions, working simultaneously to resolve the new constraint, is unknowably complex. We can't predict what form solutions take or how soon they emerge, but by acting quickly over the entire economy, we limit the chances of locking in high-carbon technology.

A carbon price is efficient. A central price signal lets the market go after the cheapest and easiest cuts first, slowly working its way to the more difficult and expensive cuts as the price rises.

A price on carbon is politically neutral. We need compromise on both sides, but nothing about a carbon price would offend any but the most ideologically entrenched. The left can lose the hubris of thinking government can impose an optimal solution and accept the priority of market-based solutions. The world is too complex for effective central planning at any level of detail. The right can abandon the absolute priority of freedom over the market and accept that the market must be tamed with a simple and effective new rule. The left get a market economy that optimizes for carbon reduction, and the right gets the market as champion. Both get a solution that works.

There are two main choices: emissions trading (i.e., cap and trade) and a carbon tax. There are also a bunch of devilish details: should it be revenue neutral or raise additional government resources? If so, why? What's the right price? How do we treat displaced industries or people who can't afford to heat their homes? My preference is a revenue-neutral carbon tax across the market, supplemented by a highly targeted cap and trade for specific industries.

Cap and trade takes two steps. First, set an emissions cap for specific industries — cement, pulp and paper, the tar sands, etc. — which lowers emissions to a long-term target.[187] Companies can invest in emissions-reducing infrastructure or choose to buy permits from competitors. The choice is economic. Those who exceed their quota buy more permits. Those who come in under their quota can sell their extra permits. Some companies will invest aggressively over the long term, betting excess permits will generate profit (the price of permits should rise as incremental cuts get harder and the target lowers). Others will take a shorter-term view and buy their way to compliance. Initial permits are often given away to generate buy-in from industry, but they should be sold for the cost of the easiest reductions. The advantage of cap and trade is we know with certainty by how much we'll reduce emissions. The downside is the cost of compliance is uncertain, since it's set by the market.

Cap and trade has its share of problems. First, the cap is politically defined and subject to the usual whims and vagaries of political expediency. Long-term targets must be firm if industry is expected to make necessary investments. An arm's length third party can be empowered to enforce targets. Second, a mere threat of cap and trade discourages investment in carbon mitigation. Why would a company invest to lower emissions if a cap is coming? Better to hold off and get credit. Early-adopters are penalized. If we're to have cap and trade in North America, we better get on with it or end the discussion. Third, it's complex, fraught with vested interests, and likely to be gamed by lobbyists and financial arbitrage.

A carbon tax is simple. There are few large-scale carbon sources in North America: coal mines, natural gas lines, refineries, and so on. Tax carbon at these sources, and the price signal will percolate throughout the economy as energy users pass on the costs. Start at a low price to avoid shocks to the system, and raise it slowly but steadily. Companies that invest in lower carbon-based energy use will get more competitive. The upside to a tax is we know the cost. The downside is we don't know by how much we'd reduce emissions. We can't know with certainty cost *and* reductions in a market economy. We set one and the market sets the other.

Both schemes can be revenue neutral. The money raised can be returned to citizens with a check in the mail or lowered income taxes. It's simple: raise prices on bad stuff and lower them on good stuff. The market delivers less bad and more good. People who choose to live a low-carbon lifestyle get wealthier, while those who choose something more carbon-heavy get poorer. But the choice is up to individuals, and companies will respond to those choices. Trillions of individual choices add up—our economy's direction will change slowly but surely. A revenue-neutral carbon price should be acceptable even to those who abhor taxes.

But should carbon taxes be revenue neutral? Huge costs are coming as we recover from storms like Sandy, mitigate economic loss from droughts, or beef up our infrastructure to adapt to the new normal. Insurance companies are already getting out of the way, so cash-strapped governments will foot most of the bill. Our principle of justice ties carbon revenue to the cost of damages. We are justified in putting carbon revenue into a large, public "stormy weather" fund, but that's unlikely to happen. From a practical standpoint, it's hard enough to get the political right to act on climate. Tying climate action to increased government revenues will be tough.

There is a compromise. An economy-wide carbon tax could be matched by reduced income tax. Select industries whose emissions are not directly linked to energy use (like the cement industry, which produces CO_2 as a core output, or the agricultural sector, whose emissions are related to farming practices) or have larger than normal profit margins (like the oil and gas sector) could implement a supplemental cap-and-trade policy targeted to their circumstance. We could then save that money for a rainy day.

Let's not forget carbon pricing will hurt lots of people. Not all purchases have a low-carbon option. The poor may have trouble heating their homes, for example. Just as we have to be cognizant of workers in declining industries, like coal mining, we have to be careful of other folks who need help during the transition. I won't tackle thorny issues of social justice here, although any analysis that doesn't is incomplete.

The price signal needs to be loud, legal, and long-term. Uncertainty prevents industries from acting, but we can make the best of existing uncertainty. The Carbon Disclosure Project (CDP) is trying to speed things up by anticipating carbon and climate risk. It argues that financial managers who ignore the threat of carbon pricing will pay dearly. Recall that fossil fuel companies will have to leave assets in the ground, and so their stock is overpriced. By telegraphing expected risk before carbon is priced, the CDP is trying to make asset managers more proactive in moving money from one sector to another.

Other policy options—feed-in tariffs (FIT), strict regulation, or renewable portfolio standards[188] — are of limited use. Each has benefits and drawbacks. Some act as training wheels for new industries, helping them get off the ground, but they can bring instability and inefficiencies to the market. While far from perfect, these more targeted policies have managed to kick clean energy into action in the absence of a carbon price.[189]

What's the right price? Principles of fairness and justice would set it to the marginal social cost of each extra metric ton (1.1 tons) of CO_2. We've seen that's a notoriously tricky calculation.[190] More pragmatically, we can work backward from a temperature target and set the price at a threshold below which all necessary cuts are profitable. The famous McKinsey Curve estimates the cost of a range of emission cuts, from the lowest (insulating your attic, a negative cost) to the highest (burying carbon in the ground). A theoretical upper limit is the cost of sucking carbon straight from the air and burying it — maybe a hundred dollars per ton of CO_2[191] — because in theory that can offset any emission source. But the sooner we act, the lower the cost will be.

Linking price to a temperature is a bit of a black art. Our best bet is to make an educated guess drawing on multi-disciplinary expertise, to be clear about our assumptions, and to refine it over time. An immediate price of forty dollars per ton might mean a two-thirds chance we stay under 2°C (3.6°F),[192] but that assumes carbon capture and storage (CCS) comes online fast. If it doesn't, the 2°C (3.6°F) target is likely out of reach no matter what the price. Carbon pricing's important, but we're so late getting started that we need a couple of turbo-boosts.

A long-term price on carbon is the most effective and efficient tool we have to build a low-carbon economy. It corrects the largest market failure of our energy system, but, in the words of Nicholas Stern, there are "a whole swath of important market failures which require us to go beyond simply a price for carbon if policy is to work."[193] Pricing carbon gets to the heart of the matter, but, because we have so little time, we'll need to tinker around the edges as well.

Accelerating the Big Three

Distributed energy is great and I'm a big fan of it, but we also need large-scale, base-load, low-carbon sources. The Big Three — next-generation nuclear (NGN), carbon capture and storage (CCS), and enhanced geothermal systems (EGS) — are too big and risky for the private sector to develop quickly. A price on carbon by itself won't do it, and they're promising enough that we can't wait. Only the public sector can unlock their promise with R&D and early support for commercial development. Right now, only CCS has that support. We need to accelerate all three then let the market pick the winner.

Publicly funded R&D is crucial, since the energy sector doesn't spend much on it.[194] Some programs exist, particularly for CCS[195] and a few for NGN,[196] but EGS lags far behind. A relatively small bet on long-term budgets for the Big Three's R&D can have big payoffs, like prior bets on aerospace, electronics, and medicine. The network of U.S. Department of Energy National Labs (Argonne, Brookhaven, etc.) is ideally situated to gather dedicated teams of experts. The intellectual property licensing strategy would open the market to multiple parties, who could then share the scale-up risk.

More important (and expensive) is early commercial support to mitigate scale-up risk and generate the shared expertise you get from learning by doing.[197] In large engineering projects like these, experience matters more than patents. The preliminary costs and associated risks create a disincentive to go first, since there's "a gain arising from the actions of those who create, develop and deploy these ideas"[198] that's shared by everyone. Pioneers are disadvantaged because competitors benefit from their risk. We bridge that gap with public-private partnerships that share risk and expertise. Let's build a dozen commercial-scale plants for each of the Big Three and see which ones grab market share from there. It's not a big ask: even if each plant cost $1 billion (which it won't, see below), that's peanuts compared to the $250 billion in energy subsidies we hand out every year to the fossil fuel sector.[199]

By next-generation nuclear I mean "breeder" reactors, not fusion ones.[200] Breeder reactors recycle existing nuclear waste, turning it from a problem into a huge energy source. Breeders burn that waste multiple times and extract at least one order of magnitude more energy than we got the first time. The final waste has a half-life measured in hundreds of years, not hundreds of thousands. Breeders also have a fundamentally different design that addresses today's safety concerns: they don't need active cooling and can't runaway into meltdown.

Will tomorrow's breeder reactors be expensive monoliths like today's fusion reactors? Time will tell, but with standardized designs, a very different safety profile, a ready source of fuel, and the added benefit of solving the seemingly intractable problem of nuclear waste, they're not a bad bet. Public fear of nuclear energy stopped development on breeders back in the 1970s, and it continues to stymie progress today. That's a mistake.

Enhanced geothermal systems (EGS) are the most promising and underfunded of the Big Three. EGS means drilling deep down (6–10 km/4–6 miles) to tap into hot, dry rock. The rock is fractured, a bit like what we do to get at shale gas now except we go deeper and extract heat instead of gas. Because EGSs are way below the water table and aren't releasing natural gas, they don't have the same environmental risks as shale gas fracking.

The fractured rock creates a loop through which water is pumped to extract heat. Water goes down and steam comes out to drive a turbine. Experimental plants in Europe and Australia have proven the principle. There's enormous energy potential, somewhere between three and thirty thousand times our energy needs—twenty-four hours a day, seven days a week. This is base-load renewable power.

EGSs can replace coal generation: drill a couple of holes next to the coal plant, fracture the rock, and replace the furnace with a heat exchanger. A 500 MW coal plant uses more than a million tons of coal per year.[201] We can get that much energy from just a couple of cubic kilometres (half a cubic mile) of underground rock.[202] If we ensure the cost of capital for the retrofit is less than the input cost of coal (plus the subsequent carbon emissions), every coal plant owner on the planet would sign up. It's doable—or at least worth the bet!

But despite the enormous upside to EGS, it remains badly underfunded. There is little, if any, support for commercial scale-up. Waterloo Global Science Initiative's (WGSI) 2011 Equinox Summit identified EGS as one of the most promising and underfunded clean-energy sources. The Equinox blueprint is Energy 2030, which recommended that a public-private consortium build ten small commercial-scale EGS plants around the world and share the resulting knowledge and expertise, such as fracking methods, how to maximize energy throughput and efficiency, and the size of the resource itself. Each plant costs about $100 million, with total funding of $1 billion—a drop in the bucket. Only when those plants operate successfully will there be enough certainty about cost and reliability to open the taps of private capital.

Carbon capture and storage (CCS) captures carbon emissions out of big smokestacks and buries them deep underground. In theory, it can make coal plants carbon-neutral. Because it promises to keep lucrative King Coal alive and well, it's the only one of the Big Three to have decent backing, including public funding for large-scale demonstrations. We need to show how it works in different geographies and geological formations, with lots of industrial partners to spread the knowledge. For commercial viability, it needs a carbon price of between thirty dollars and sixty dollars per ton of CO_2 (adding 3 to 6 cents/kilowatt hour to your electricity bill).

It's important for the coal sector to see a path forward to the low-carbon economy. It's better to keep this sector engaged instead of defensive, but I think traditional CCS is just a distraction to buy time. The promise of carbon-capture-ready plants provides political cover and it doesn't amount to much for the plant itself, maybe an extra valve or two, but we saw in Chapter 5 just how much additional infrastructure it needs to work.

There's a more promising route: capture the CO_2 by spraying the stack gas with something like carbonic acid mixed with nickel nanoparticles (as a catalyst to speed up the reaction). You end up with a giant pile of chalk, which can be used as a commodity for building materials or whatever. Less macho, but way more effective!

Traditional CCS needs competition. A dozen operating plants of NGN and EGS might even the playing field. I don't think CCS can beat either of them in a mature and competitive environment, but right now CCS is the only big bet we're making. That's a mistake. Let's accelerate all three.

Capital: From Defense to Offense

Twenty years ago, a price on carbon would have been enough to keep us from going over the climate cliff, but not today. We need to turbo-boost the flow of capital so it plays offense on climate, not defense. Defense avoids risk. Offense takes risks to solve a problem. Defense might avoid investment in a coal plant, while offense might seek investment in a geothermal plant. That's a subtle but important difference.

We can force capital to take the offense on climate by issuing green bonds, which are like war bonds for the environment.[203]

An asset manager's duty is to protect the financial interests of clients, not to solve global problems. Big capital's first reaction to carbon and climate risk is defensive: protect assets by divesting from problem companies and sectors. Capital flows out of coal and drought-prone agriculture in the American south . . . into what? Offshore wind or the Big Three? New venture funding for cleantech? Not likely. Big capital is conservative and likes to repeat past successes. Well-established sectors, such as solar farms and onshore wind, might see an uptick, but new energy technologies—like the Big Three or next-generation solar—will languish because they're seen as risky. The effect of a carbon price on energy infrastructure will be gradual and mainly indirect. Green bonds give asset managers a reason to put some all-stars on the climate offense.

The idea is simple and leverages the public *and* private sectors. Citizens and asset managers buy a green bond, which is backed by their national government. The private sector bids on the right to manage the money, which goes into a green fund to provide low-cost financing for new low-carbon infrastructure. Government's role is limited to guaranteeing the money and providing one key metric of success to the private sector managers: reduce the most carbon at the lowest cost. The only cost to government is defaulted loans.

Those in the private sector are given an incentive to use their creative expertise to maximize carbon reduction and minimize cost. Their fees and bonuses are linked to the performance of the loans. No defaults means no cost to government and big bonuses in the private sector. Lots of defaults mean no bonuses. The fund managers will diligently pick and oversee the teams, technologies, and projects for which they lend. They might put liens on equipment and power contracts in case of default. They might ask for matching private funds. Green bonds combine advantages of the private sector markets with the low interest rate of the public.

The private sector picks the projects, although it could be forced to favor underfunded next-generation technologies like the Big Three. To ensure the green fund doesn't interfere with existing capital flows, projects are categorized according to the difference in available private and green fund interest rates. The green fund won't be able to participate below a certain threshold because the private sector is already there. This would skew the fund toward enabling those more promising next-generation technologies that still contain too much risk for the conservative nature of traditional big capital.

The green bond can pay returns because energy projects generate revenue. These aren't bridges, roads, or sewers. Any carbon credits they produce can be used to juice returns, attracting more capital. Carbon credits return to project owners only when they pay the loans back, further reducing loan losses. The private fund managers are paid with the spread between the bond interest and the project lending rates.

A Canadian green fund can add a new twist to the perennial federal-provincial dance on energy strategy. In Canada, energy remains a provincial jurisdiction. A national green fund can be the carrot that makes provinces compete to create a positive economic climate for clean-energy infrastructure. The private sector fund managers pick projects with the least risk, so provinces that reduce risk the most will win the lion's share of the capital. Since it's the private sector that makes the decision, the federal government remains one step away and can duck blame for picking favorites.

A global green fund backed by the World Bank or IMF can be tied to national targets in a similar way. Countries compete to reduce the risk for next-generation energy infrastructure. Total fund capital is tied to global emission targets. With carbon credits juicing returns and government backing lowering risk, there should be no trouble raising whatever capital we need to jump-start climate action — particularly in today's risk-adverse credit climate. Tweaking bond interest rates can open huge flows of capital where we need it most.

We might even provide fossil fuel companies, whose balance sheets are decimated by a carbon price that forces them to leave assets in the ground, with low-cost capital to redeploy their engineering, market presence, and know-how into low-carbon energy. This might feels a bit like bribing the bad guys — do they really need the money? But which would you prefer, co-opting such a powerful economic force or fighting it?

Pick Your Targets

There is low-hanging fruit on the carbon tree. Let's pick it. But there are other bits we'll never reach (or might not even want to). Let's not waste our efforts there. What follows is a miscellany of targets.

We'll never go without long-distance air travel, for example. Short-haul flights can be replaced with high-speed rail, and lots of business trips are now redundant. Personal travel, however, is part of the glue that makes us a global community, helping to form our shared humanity. The backpacking trips of young adults seeking to explore the world around them, the cultural journeys we make as adults, and even the "love miles" to visit friends and relatives — that carbon burn is not our best early target; we gain too much in so many ways compared with other, more easily replaced emissions.

That doesn't mean we shouldn't make an effort. We can prioritize bio-based aviation fuel over other transport fuels because most other forms of transport can go on a low-carbon diet without much sacrifice. We can increase the efficiency of our aviation fleet. A lot of business travel can be replaced with a good telepresence network, but let's not pretend we're going to stop flying or that trying would even be a good thing.

An aviation carbon diet can be finessed to preserve the irreplaceable value of personal travel. Perhaps we could each have an annual allowance of no-carbon-tax personal travel, after which a high price kicks in. We would start at the higher price for business travel. If the trip is worth taking, businesses will find a way to way to pay for it. If it's not, there are other ways to do the deal. Flights generating the shared cultural experiences that knit us together as a global community can be the last ones we seek to curb.

In our cities, buildings account for nearly three-quarters of carbon emissions. They're some of our lowest-hanging fruit. I developed a building in downtown Toronto — Planet Traveler — which I claim is North America's lowest-carbon hotel.[204] We put geothermal piping underneath a city-owned laneway, showing it can be done in a dense downtown core.[205] We have solar thermal power for hot water, a bit of solar PV (which doubles as a shade awning on our awesome rooftop deck), and capture waste heat before it goes down the drain. With super-efficient LED lighting, we can light up the entire building like a Christmas tree — inside and out — for less energy than what a four-slice toaster uses.

Leveraging just five percent of the building's capital, we reduced our energy use by three-quarters, and we did it at a profit and lost no creature comforts for our guests. Even better, the loan payments on the retrofit are less than the energy savings, so we've been cash flow positive from day one. We are wealthier as hoteliers for making those emission cuts, not poorer. That rule of thumb — five percent of capital to unlock a three-quarters cut — applies to commercial buildings of almost any size.

Not all existing building owners pay the energy bill, so they're less motivated to make changes. That's where an ESCO (an energy services company) comes in. They foot the bill and share the benefits.

Everyone's happy: the tenants get a workplace they can be proud of, the owners make more money, and so does the ESCO. In today's crammed cities, even the utility that sells electricity is happy because they can better manage an overheated grid.

Planet Traveler can be part of a Canadian economic good news story. Canadian engineers planned the work, and Canadian trades-people carried it out: plumbers, electricians, carpenters, and drillers. The hotel even has Canadian LED lighting and solar thermal panels. The solar PV racking is Canadian-made, and so is the drain-water-heat-recapture equipment, and so on. It would be the same for any country—the economic benefits of building retrofits are largely local. Here's a stimulus idea: fund a retrofit of your entire country's built environment. Every building owner gets richer, and it creates domestic jobs and makes the entire economy more efficient. Where do we get the money? How about city-issued green bonds to provide funding to ESCOs who do the work, take the risk, hire the workers, etc.

If every building did what we did—and most could—Canada would zoom past our abandoned Kyoto promises! To hear Canada's government talk about it, hitting those Kyoto targets would render us uncompetitive, poorer. Absolute nonsense. At Planet Traveler, we cut our carbon by three-quarters, and we're richer, not poorer, for doing it. Buildings account for forty percent of our carbon emissions. Kyoto required a six percent cut. You can do the math. It's too late for Kyoto now, but that doesn't mean we can't start.

For new buildings, we can change the building code. Developers have to build new condos as competitively as possible. That means skipping the higher capital costs of geothermal heating and cooling, for example, unless condo buyers get savvy enough to calculate the net present value of the energy savings and add it to the price of the condo. Unlikely! It's easy to sell granite countertop upgrades, but geo-thermal or drain-water heat recapture—well, it's just not as sexy. The construction industry is conservative and competes on price. It's difficult to move first. A building code that demands something close to what we've done at Planet Traveler would change the industry overnight and keep an even playing field for developers.

The less businesses spend on energy, the more they can spend in productive ways, such as innovation, market development, and expansion. Japan and Germany generate double the economic output of North America per unit of energy. That's not because they don't have heavy industry. It's because they've invested in the capital infrastructure that reduces energy inputs. Their economies are more efficient and more competitive as a result. The good news for North America is that we've got some really low-hanging fruit that we can pick.

Opponents of Canada's tar sands will tell you it's one of the world's most carbon-intensive and environmentally destructive sources of oil. They're right. They're also right to point out that Canada has no realistic hope of meeting national commitments on emissions reductions if we continue to develop tar sands at the current rate. Proponents will point out that it's a politically stable source of oil in a world fraught with risk. They're also right. The tar sands are one of Canada's most divisive subjects. Whether environmentalists like it or not, current development will not be shut down, but whether the oil sector likes it or not, their social license to expand operations is under legitimate threat.

The tar sands have been developed at breakneck speed, with nearly open-ended expansion rights handed to the oil giants. It's become difficult to get the oil to market, and as a result the oil is sold at a significant discount.[206] Oil companies need more pipelines. That's where the battles are being fought: big oil companies backed by an aggressive Harper government versus embattled environmental groups, a growing portion of the public, and Aboriginal Canadians. Opponents have thus far successfully blocked the Northern Gateway pipeline through British Columbia to the west coast and the doubling of flows through Keystone XL into the U.S.

I propose the following trade-off: we accept additional oil flows going south on Keystone in exchange for a hard cap on future pipeline development and priority access on the new pipeline for Canadian biofuels. Like all good compromises, everyone loses something. Carbon hawks (like me) swallow more high-carbon infrastructure, the oil sector accepts limits on development, and everyone benefits from the new support for Canada's massive biofuels potential. We also take pressure off the Northern Gateway, which is off the charts when it comes to threatening pristine wilderness and aggravating Native sensibilities. The compromise emphasizes two pragmatic facts: we're not shutting down what's already developed (at least for some time), and the oil sector admits to limits on future development.

What's the Cost?

How much will it cost to turn down the thermostat?

The cost depends on how fast and deep we cut emissions. The faster we act, the cheaper it is. The deeper we go, the more expensive it gets. But let's not beat around the bush: it's a huge job and will cost lots of money — trillions of dollars. But relative to what we already spend on fossil fuels, to the size of the economy, to the time it takes to spend it, to what we get in return, and, most importantly, to what happens if we don't spend it — that's actually not much.

Typical climate cost and benefit analysis is a mug's game. Recall proponents (and critics) try to justify the cost of action by calculating the benefits. We might save some of our agriculture, forestry, and fisheries industries, for example, or reduce damage from hurricanes, drought, and floods. What's that worth? To answer, we use models loaded with assumptions like the discount rate, damage function, and dumbed-down climate science. Forget that crap. Let's pretend there are *no benefits* and *only costs*.

Let's skip the trap of putting a dollar value on our ecosystem. We can call the benefit saving ourselves and our way of life and peg its value at zero dollars. Let's drop all claims to economic benefit. Forget trying to measure the benefit of saving our fisheries or agriculture and avoiding mega-storms and droughts. Let's also set aside the economic stimulus from a redistributed carbon tax and lowered imported energy bills as well as the health benefits from cleaner air. Let's also assume there's no upside for energy security. If we can accept the worst case, the economic arguments against action are irrelevant.

We're buying climate insurance. It gets us a temperature target. The cost is incremental costs over business as usual. We can express it in dollars per ton of carbon, or we can put it in absolute terms — the language of "trillions of dollars" that critics often use to scare us into inaction. Let's use that language, without taking any benefits into account, and see how bad it really is.

The IEA has four scenarios representing pathways to 2035, each with a different level of warming. Our climate insurance is the cost of subsidies for clean energy (direct and indirect) plus the carbon price (this double-counts some costs because carbon costs can fund the subsidies — again, the worst case).

The Current Policies Scenario is business as usual. We don't do much of anything and hit 6°C (10.8°F).[207] Bleak! For perspective, we still need to invest nearly $38 trillion between now and 2035 to meet growing demand, most of it on oil and gas.

In the New Policies Scenario we make a bit of effort by supporting renewables with $4 trillion in additional subsidies.[208] There's carbon pricing in China and a few other countries. The temperature target drops to 3.6°C (6.5°F). Still bleak!

The most aggressive scenario is the 450 Scenario. Emissions peak by 2020. We get maybe a fifty-fifty chance of limiting warming to 2°C (3.6°F). Better! That requires an additional $16 trillion in investment and more aggressive carbon pricing. This is starting to look expensive: that's $20 trillion of extra investment, and something like another $20 trillion in carbon costs.[209] Wow, maybe this is a bad deal.

Hang on a moment. The 450 Scenario includes the Efficient World Scenario, in which aggressive investments in energy efficiency cost nearly $12 trillion but save $17.5 trillion in reduced energy use and another $6 trillion in avoided supply infrastructure. We make $11.5 trillion. It's Planet Traveler on steroids! There's more. We could add the $18 trillion in increased GDP gotten from more efficient allocation of resources[210] and benefits from lower fuel costs due to lower demand and . . . But we're looking at the worst-case—let's not count all that extra goodness.

The final bill—no economic benefits included—drops to $30 trillion. Call it an even trillion a year. There are seven billion of us, so that's $142 per year, per person. Climate insurance that gives us a decent chance at saving civilization comes in at three bucks a week for every man, woman, and child on the planet. A weekly coffee and donut to save us from ourselves. Maybe people in rich countries have to buy a few extra coffees because the poor can't afford them. Is that really such a bad deal?

We've overstated the cost, of course, because we've pretended there are no benefits, but there are benefits, lots of them. Somebody's going to invent those new technologies, make all that equipment, build those power plants, retrofit those buildings, engineer a new grid, export technology to India and China. That's a lot of upside, and we haven't included the $18 trillion from more efficient use of our resources or the benefits that are so hard to calculate: fewer crazy storms, less droughts and wildfires, lower costs of adaptation, energy security, and lowered costs of militarizing the Middle East. Maybe we'll even save the global fishery. The global economy needs a shot in the arm, and rebuilding our energy infrastructure can do it.

Look at it another way. The IEA assumes a 3.5 percent annual growth in the global economy, from $70 trillion now to $165 trillion by 2035. The cost of our insurance is being delayed a few years before doubling our global wealth. The cost of going without insurance is the near-certainty our economy shatters.

There is no cost to dialing down the thermostat. There is only benefit.

Policy Approaches

It doesn't take sophisticated knowledge of negotiation dynamics to know the consensus-based United Nations Framework Convention on Climate Change (UNFCC) series of conferences (Conference of the Parties or COPs) was flawed from the outset. In fact, the impossibility of getting hundreds of nations to reach a consensus is obvious enough one wonders if it wasn't set up to fail. The COP process is flawed because the most recalcitrant country drags the rest of us down with it. That country is often the United States. It's a race to the bottom.

The initial Kyoto Protocol got hammered through mainly because the developing world didn't have to commit to doing much of anything, and the sanctions for the developed world were toothless (much as the U.S. Senate railed against loss of sovereignty[211] and Canada bemoaned financial ruin[212]). The bad news is the Kyoto Protocol passed only because it's nothing more than a good-natured agreement to try. The good news is most of the countries that took their commitments seriously from the start have (or are very close to) meeting those targets. Kyoto showed we *can* get the job done if we put our minds to it.

Getting past Kyoto to deep cuts and real sanctions won't happen at the ever more embarrassing distraction the COP process has become. That's not to say it hasn't attracted well-intentioned, deeply committed people: they're filled with the best and brightest the climate movement has to offer. But a process requiring unanimity on even the most benign sort of "agreement to agree" language is never going to get anywhere. And when countries like China—who have trillions in cash reserves—cry poverty and demand a continued transfer of wealth from developed countries, it further hijacks the process with a demonstration of bad faith.

The tragedy of the COP process is that it's distracted a lot of genuinely committed people for so long. It also fooled decent citizens into thinking something substantial was happening. As a result, it's been a perfect straw man for the fossil fuel industry.

We need a new framework. We need to ditch unanimity in favor of strong bilateral and multilateral agreements based on cold, hard mutual economic benefit. While Europe has shown great moral leadership on emissions,[213] it's naive to think the U.S., China, or India will act for altruistic reasons. New agreements can be based on the realistic assessment that, sooner or later, the world will de-carbon. The economic opportunity de-carboning offers is stupendous—the clean-energy market is projected to be between $2 trillion and $3 trillion annually by 2020. Economies that position themselves intelligently will benefit, and those who ignore the opportunity will not. It's just a matter of time.

For example, the United States and China each bring unique strengths to the climate fight. The U.S. remains unmatched in developing new technology and leveraging expertise in venture capital to bring it to market. China dominates low-cost, high-volume manufacturing (and brings the balance sheet of the China National Bank). A U.S.-China agreement might see China exchange unhindered access to next-generation technology for respect of American intellectual property, complete with requisite licensing fees. Both could agree to price carbon on domestically generated energy, which creates a huge initial market. In essence, America invents, China manufactures, and both purchase. That's a partnership strong enough to dominate low-carbon technology production and lead to huge export markets elsewhere.

Other countries will have no choice but to respond and will scramble to define their own role. Soon it will be a race to the top instead of a race to the bottom. To ensure *all* countries eventually join the fight, we could modify existing institutions like the World Trade Organization (WTO) to create a level playing field. Countries that refuse to price carbon in domestic energy markets will face harsh carbon tariffs from their trading partners.

This is classic game theory—a mix of carrots and sticks, with the strongest players making the first moves based on mutual economic benefit. Countries benefit if they choose to participate in the largest global market of the early twenty-first century, but they risk getting hurt if they refuse. Carrots and sticks are better than just sticks, particularly unilaterally imposed sticks. President Obama, Premier Li, the next move is yours.

Choosing a Future

Sceptics have come up with lots of arguments for why we shouldn't act on climate. None stand up to scrutiny. Some claim the science is uncertain. That's just untrue, and what uncertainty remains renders climate disruption more dangerous, not less. Others worry the economic dislocations (like workers in the coal industry) will be too painful. It's true that there will be winners and losers. Real people with real families will get hurt in this transition. We'll have to take care of those workers as best we can, but the car put the horse and buggy out of business. Progress can't be held hostage by the vested interests of the few over the benefits (or even survival) of the many.

Lots of arguments are about cost. We've seen those arguments are also wrong. Rebuilding our energy infrastructure can stimulate the economic malaise of the developed world. The hard truth is that there is no longer such a thing as high-carbon economic growth. We might squeeze out another decade of growth, but the carbon we burn to do it will more than reverse those gains.

There are lots of things we can reasonably disagree on when it comes to climate disruption, but acting quickly and determinedly is not one of them. Our public debate is much better spent on the complex problem of how best to act. How much effort on adaptation versus mitigation? Do we embrace nuclear? How to treat emissions from the developing world? Who pays for it all and how? What will rising fossil fuel prices mean for the poor? What's the proper role of national governments? The U.N.? Can't we just geo-engineer our way out of the problem? What's the best way to ensure all countries step up to the plate?

Climate disruption, and all its attendant economic, political, and social Gordian knots, is the most complex issue we've ever faced. Any of the solutions I've presented can be debated in good faith, but one thing we must agree on is the urgent need to act. Disagreement about the existence or urgency of the problem does a deep disservice to the generations who came before us. Our civilization was built brick by brick by countless generations. We are the descendants of the people who got us out of the muck, through the Enlightenment, and into the twenty-first century, with its unparalleled wealth and knowledge. It is an insult to our collective and historical intelligence to continue to ignore the warming climate and all its repercussions.

Humankind has come a long way in the last couple of thousand years. From Rome and the birth of Christ through to our wonderfully complex global economy, we stand on the shoulders of giants. Art, literature, science, culture, and our civic structures, all are results of our long journey to the present. Who knows what further adventures might await? We'll only have the chance to find out if we manage to squeak through the climate crisis and stop our mad gallop toward the climate cliff.

Some believe our destiny is set, perhaps because we are biologically determined to act as all creatures do and bump up against the ceiling of our habitat (ours happens to be the entire planet), perhaps because we have fallen in the trap of outdated ideas and are already too deeply committed to an unsustainable path. I don't believe either argument to be correct, and it's not just because I don't like such a bleak, determinist view. That's just wishful thinking. I'm convinced both views are incomplete.

Existentialists, faced with the despair of a dark human nature laid bare in the horrors of the Second World War, realized that the future they faced was neither determined nor bleak—but ours for the choosing. We are profoundly free because we, uniquely among creatures, have the capacity to choose our future. We can choose one of sustained abundance, but we must *make* that choice. Together, we must *demand* that future. We can commit to a vision of ourselves and of our relationship with this planet far advanced from that which we have today. We are not masters of our domain unless we are first masters of ourselves.

SUGGESTED READING

Tim Flannery, *Here on Earth: A Natural History of the Planet* (2010)

Peter Newell and Matthew Paterson, *Climate Capitalism: Global Warming and the Transformation of the Global Economy* (2010)

James Gustave Speth, *The Bridge at the End of the World: Capitalism, the Environment, and Crossing from Crisis to Sustainability* (2008)

Nicholas Stern, *The Global Deal: Climate Change and the Creation of a New Era of Progress and Prosperity* (2009)

Endnotes

Introduction: Frogs, Hot Water, and Us

1   See *World Energy Outlook 2011* and *World Energy Outlook 2012*, International
    Energy Agency.

2   I have written elsewhere of technology. See *Kick the Fossil Fuel Habit: 10
    Clean Technologies to Save Our World* (EcoTen Publishing, 2010).

3   China is an obvious exception, and the degree of real democracy in places
    like Russia (and, for some critics, even the U.S.) is questionable. That said,
    I'm limiting my target to the broad economic structure in which I operate.

4   There have been a number of joint statements signed by equivalents of
    the U.S. National Academy of Sciences, including the Science Council
    of Japan, Indian National Science Academy, Chinese Academy of Scienc-
    es, Royal Society, Russian Academy of Sciences, Academie des Sciences,
    Royal Society of Canada, and on and on. The point I'm making is simple:

at the highest levels of scientific inquiry, the view on climate change is unambiguous and collective.

Chapter 1: From Serious Science to the Theater of the Absurd

5   Revkin, Andrew C. "Politics Reasserts Itself in the Debate Over Climate Change and Its Hazards," *New York Times*, August 5, 2003.

6   Bush, G.W., "Text of a Letter from the President to Senators Hagel, Helms, Craig, and Roberts," *White House Archives*, March 13, 2001.

7   See "In President's Words: 'A Leadership Role on the Issue of Climate Change,'" *New York Times*, June 12, 2001.

8   As reported in "Harper's letter dismisses Kyoto as 'socialist scheme,'" *CBC News* online, January 30, 2007.

9   See "Tony Abbott makes a 'blood pledge' to repeal carbon tax after it passes lower house," *The Australian*, October 12, 2011.

10   See "Tony Abbott mounts a powerful argument against Gillard's carbon tax," *The Australian*, July 9, 2011.

11   See "Harnessing collective intelligence to address climate change: The Climate Collaboratorium Copenhagen Challenge" on *ClimatecoLab.org*.

12   D. Tapscott and A. Williams, "Macrowikinomics: Opening the Kimono on Climate Change," *Huffington Post* (2010).

Chapter 2: The Siren Song of Denial

13   Here I include institutions, their funders, and some pundits and politicians: the Heartland Institute, the Competitive Enterprise Institute, Exxon Mobile, Glenn Beck, Mitt Romney, and even Stephen Harper before he became Canada's prime minister.

14   The name has been changed but the conversation was real.

15   I edited only for coherence and length. I have stayed as close to the con-
      ceptual mapping as possible. Much of it was by email, and so there was a
      written record.

16   International Panel on Climate Change.

17   "Climategate" was a set of hacked emails from the University of East Anglia
      that purported to show climate disruption is false. Multiple independent
      investigations (including one by *The Economist*) showed malfeasance might
      have occurred, but this has no bearing on the vast preponderance of evidence
      for climate disruption. It was a tempest in a teacup, leveraged for propagan-
      dist purposes.

18   What I'd give to debate Rush or Glenn or Wente or Murphy or . . . any
      takers?

19   These are the regular patterns of the Earth's wobble around its axis of orbit,
      and rotation around the sun, which give rise to long-term variations in our
      planet's climate.

20   Gallup 2008 environment poll.

21   K.M. Norgaard, *Living in Denial* (MIT Press, 2011), xix.

22   Norgaard, *Living in Denial*, xix.

23   M. Rosenberg, "Self-Processes and Emotional Experiences," in *The Self-
      Society Dynamic: Cognition, Emotion and Action*, ed. Judith Howard and
      Peter Callero (Cambridge University Press, 1991), 123–142.

24   R. Lifton, *The Protean Self: Human Resilience in an Age of Fragmentation*
      (New York: Basic Books, 1993). See also *Hiroshima in America: Fifty Years of
      Denial* (G.P. Putnam Sons, 1995).

25   "Although Koch Industries intentionally stays out of the public eye, it is now
      playing a quiet but dominant role in a high-profile national policy debate on
      global warming. Koch Industries have become a financial kingpin of climate
      science denial and clean-energy opposition. This private, out-of-sight corpo-
      ration is now a partner to Exxon Mobil, the American Petroleum Institute,
      and other donors that support organizations and front-groups opposing pro-
      gressive clean-energy and climate policy. In fact, Koch has outspent Exxon

Mobil in funding these groups in recent years. From 2005 to 2008, Exxon Mobil spent $8.9 million while the Koch Industries–controlled foundations contributed $24.9 million in funding to organizations of the climate denial machine." T. Carrk, *The Koch Brothers, What You Need to Know About the Financiers of the Radical Right* (Center for American Progress Action Fund, April 2011).

26 Schwartz, P., & Randall, D., "An Abrupt Climate Change Scenario and Its Implication for United States National Security," *Pentagon Report*, October 2003 (available at http://web.mit.edu/hemisphere/events/Pentagon_Warming.pdf).

27 Not all religious beliefs are benign. Belief in the Rapture (a day when believers are sucked into the clouds to watch the rest of us burn on Earth), for example, makes it difficult or even impossible to care about climate disruption. It might even be a good thing, a prelude to the Rapture. Other religious views are troubling: we cannot possibly harm God's Earth, climate disruption is payment for a nation of sinners, and so on.

28 These are just two from a treasure trove of examples in cognitive scientist and Nobel laureate Daniel Kahneman's brilliant book *Thinking, Fast and Slow* (Farrar, Straus and Giroux, 2011).

29 As reported by Kahneman in *Thinking, Fast and Slow*, 82, but originally due to an experiment by psychologist Solomon Asch.

30 G. Lakoff, *The Political Mind* (Penguin, 2009), 3.

31 P. Kellstedt, S. Zahran, and A. Vedlitz, "Personal Efficacy, the Information Environment and Attitudes Toward Global Warming and Climate Change in the United States," *Risk Analysis* vol. 28 (2008): 113–126.

32 For a full exploration of human abilities that cannot be captured by computers, see H. Dreyfus, *What Computers Still Can't Do* (MIT Press, 1999).

33 For details on this and other examples of open-ended complexity that only humans can cope with see T. Rand, *Intuition in Philosophical Analysis: Taking Connectionism Seriously* (University of Toronto, Doctoral Thesis, 2001).

34 Some argue it's actually infinite (because of a kind of regress) and some

only practically infinite. See Rand, *Intuition in Philosophical Analysis* for my own view.

35   This is an old philosophical chestnut: you are to imagine a disembodied brain grown in a vat. Can that brain think or be conscious?

36   The inner working of neural networks, and the way they discern patterns and form thoughts, was the subject of my own academic work; see Rand, *Intuition in Philosophical Analysis*.

37   In testimony to the U.S. Congress, as transcribed by PBS Newshour, October 23, 2008.

38   These biases are well understood and relatively uncontroversial. While I reference Daniel Kahneman throughout this section, any number of modern cognitive science texts agree.

39   Kahneman, *Thinking, Fast and Slow*, 97.

40   Kahneman, *Thinking, Fast and Slow*, 97.

41   Kahneman, *Thinking, Fast and Slow*, 81.

42   Kahneman, *Thinking, Fast and Slow*, 103.

43   Kahneman, *Thinking, Fast and Slow*, 82.

44   Kahneman, *Thinking, Fast and Slow*, 284.

45   Kahneman, *Thinking, Fast and Slow*, 305.

46   Kahneman, *Thinking, Fast and Slow*, 90.

47   M. Heffernan, *Willful Blindness* (Anchor Canada, 2011), 6.

48   Orwell, George, *Nineteen Eighty-Four* (Martin Secker & Warburg Ltd, London, 1949), 32.

49   Naomi Klein, "Capitalism vs. Climate," *The Nation*, November 28, 2011.

50   C.S. Lewis, *Selected Literary Essays* (Cambridge University Press, 1979), 265.

51   I. McEwan, "Pessimism Is a Luxury We Can't Afford," *The Guardian*, April 22, 2005.

52   See T. Essig, "Climategate-gate: The Dangerous Psychology of Ongoing Climate Change Denial," *Forbes*, November 26, 2011.

CHAPTER 3: COMPLEXITY AND THE MYTH OF THE FREE MARKET

53   This unrestrained fervor was kicked off in earnest by the repeal of the Glass-Seagall Act by the Clinton administration.

54   It was not just the open window through which the Federal Reserve shoveled cash at near-zero cost to the banks, but the fact that many of those banks in turn lent that money straight back to the Treasury at higher rates by buying U.S. Treasury bonds. The banks arbitraged the federal government's own spread between the cost of printing cash and borrowing back the money. The cost of these subsidies (effectively zero borrowing costs) is estimated to be $120 billion a year in the U.S. alone. See B. Milner, "Lack of Monetary Rigour Fuels Permabears Gloomy Outlook," *The Globe and Mail*, March 5, 2012.

55   For a highly readable and well-researched book on the duplicity and bald dishonestly of the entire banking system—from Alan Greenspan through Goldman Sachs right down to the sub-prime salespeople—see the wonderful book *Griftopia* by Matt Taibbi (Random House, 2010).

56   See John Perkins, *Confessions of an Economic Hit Man* (Penguin, 2006) for a fascinating insider's view of the history of national debt in developing countries.

57   See Joseph Stiglitz, *Globalization and Its Discontents* (Norton, 2003), by the ex-head of the World Bank, for a full and readable account.

58   Damage estimates of a Category 4 or 5 hurricane hitting Tampa run between $50 to $65 billion (see M. Hertsgaard, *Hot: Living Through the Next Fifty Years on Earth* (Houghton Mifflin Harcourt, 2011, 147). The Florida budget is around $60 billion.

59   In 2002, the U.S. Congress passed the Terrorism Risk Insurance Act, making $100 billion available as a backstop to private insurance for places like Chicago's Sears Tower and San Francisco's Golden Gate Bridge.

60   This is not a comprehensive accounting of the role the public sector has played in wealth creation, but a quick overview. For a detailed analysis, I

would recommend Michael Lind's *Land of Promise* (HarperCollins, 2012).

61 See Paul Krugman of the *New York Times* as an advocate for strong stimulus in the current global recession, as opposed to David Cameron's Conservative British government, who have instead chosen austerity.

62 Stiglitz is the ex-chairman of Bill Clinton's Council of Economic Advisors and ex-president of the World Bank.

63 Stiglitz, *Globalization and Its Discontents*, 22.

64 The U.S. government also protected the automotive industry from foreign competition, by exempting large gas-guzzling trucks from fuel-economy standards while putting up tariffs against equivalent gas-guzzling imports. Arguably, this protectionism helped create the crisis that required the recent bailout. Government interference, in and of itself, can be good or bad.

65 For an idea of the kind of commitment a government needs to make to enable a growing fossil fuel sector, in terms of providing stability, access to resources and even diplomatic arm-twisting, see Daniel Yergin's two books, *The Quest: Energy Security and the Remaking of the Modern World* (Penguin Press, 2011) and *The Prize* (Simon and Schuster, 2009).

66 This is a vast oversimplification. There are many different historical and conceptual interpretations of economic theory, as well as more subtle derivations from the core theory, but the key points hold and appear in much the same form in many standard economic textbooks. For those who would like a more fleshed-out story, including an overview of market complexity, I recommend Eric Beinhocker, *The Origin of Wealth* (Harvard Business School Press, 2006).

67 But more to the point, it seems so naive as to be almost quaint. I'm not sure child-laborers in South America would be able to judge the fairness of their daily exchange.

68 A. Smith, *The Wealth of Nations*, chap. 2, bk 1 (1776), 15.

69 See "The Greenspan Effect: The Doctrine Was Not to Have One," *New York Times*, March 18, 2010.

70 J.M. Keynes, chap. 3 in *A Tract on Monetary Reform* (Macmillan and Co., 1923).

71   Beinhocker, *The Origin of Wealth*, 23.

72   E. Andrews, "'Maestro' Leaves Stellar Record and Murky Legacy," *International Herald Tribune*, August 26, 2005.

73   Complex systems theory describes how complicated but orderly systems change over time. It's a field that inherited much of the mathematics of chaos theory in the 1980s. From the biosphere to ant colonies to social behavior on the Internet, complex systems operate far from a classic, physical equilibrium since they need a constant input of energy to generate interesting order.

74   This is obviously a topic that deserves a book of its own. This section is a vast oversimplification that brings out a few key points. Some view the evolutionary component as analogous, some as quite literal. For further reading see Stuart Kauffman's work, including *Reinventing the Sacred* (Basic Books, 2008) as a theoretical basis, and Beinhocker, *The Origin of Wealth* for a practical overview of the field.

75   S. Kaufman, *Reinventing the Sacred*, 153.

76   http://bfi.uchicago.edu/about/tribute/mfquotes.shtml

Chapter 4: Economics: This Dismal Science

77   Lomborg is much vilified in the scientific community, and he was famously hit in the face with a pie by an angry climate scientist. The vitriol is highest among hard-core scientists, mainly because his understanding of climate science is rudimentary at best and because he's fast-and-loose with the truth and often outright wrong.

78   A discount rate is a technical term that defines the amount by which we reduce the value of a dollar in the future. Since the economy grows and a dollar invested today gives more than a dollar in the future, today's dollar is "discounted." How to set the discount rate is highly controversial because it can have a huge effect on the analysis.

79   B. Lomborg, *Cool It* (Vintage, 2010), 22–23.

80 That this discussion focuses on Kyoto, rather than some post-Kyoto treaty, is not to the point since my general criticisms will apply to any policy subject to the DICE analysis.

81 Lomborg, *Cool It*, 33.

82 Lomborg, *Cool It*, 32.

83 Lomborg, *Cool It*, 33.

84 Lomborg, *Cool It*, 37.

85 Lomborg, *Cool It*, 160.

86 Lomborg, *Cool It*, 138.

87 Lomborg, *Cool It*, 12.

88 Lomborg, *Cool It*, 19–20.

89 This is really odd, considering Lomborg's background is in statistics. What this reveals, in my view, is both how seductive a worldview can be and, perhaps, how disingenuous one might become when defending it. .

90 W. Nordhaus, ed. *Economics and Policy Issues in Climate Change* (Resources for the Future, 1998), 18.

91 By "complex" I mean there are a large number of mutually interacting sub-disciplines that together form the planet-sized puzzle of climate science: molecular physics, atmospheric chemistry, oceanic and atmospheric circulation patterns, geology, biofeedback loops including everything from plankton to soil to tropical forests, oceanic-atmospheric interactions — the list goes on. By "non-linear" I mean emergent behavior that's highly unpredictable, subject to massive and rapid changes, and which may have many points of stability.

92 This does appear to be the goal. Nordhaus states that the philosophy behind simplifying the science is so "the optimization model is empirically and computationally tractable." *A Question of Balance* (Yale University Press, 2008), 43. This means we're able to stuff it into a spreadsheet. He also admits that "no definitive answers are possible, given the inherent uncertainties — rather . . . these models strive to make sure answers are at least internally consistent," which means we seem to be more worried about the model looking pretty than it giving accurate answers.

93 "A recent estimate suggests that the perennially frozen ground known as permafrost, which underlies nearly a quarter of the Northern Hemisphere, contains twice as much carbon as the entire atmosphere," "As Permafrost Thaws, Scientists Study the Risks" *New York Times*, December 16, 2011.

94 The spirit of this analogy can be attributed to Wallace Broecker, who famously said, "The climate system is an angry beast and we are poking it with sticks" (as quoted in M. Weitzman, "GHG Targets as Insurance Against Catastrophic Climate Damages").

95 My use of tar sands instead of oil sands will, to some, indicate a biased position. Perhaps. But it is a more accurate term: we are melting the equivalent of tar and upgrading it to oil with the addition of hydrogen. It is also the original term; oil sands is the result of a re-branding campaign.

96 In an almost comic note, Nordhaus adds, "unexpected global changes are inherently difficult to predict." Yes, the unexpected is difficult to predict. This is known as a tautology — or stating the obvious.

97 Nordhaus, *A Question of Balance*, 28.

98 Nordhaus, *A Question of Balance*, 28.

99 Nordhaus, *A Question of Balance*, 27.

100 See M. Weitzman *Fat-Tailed Uncertainty and GHG Targets* in Bibliography.

101 See Sherwood & Huber, 2010 in Bibliography.

102 This is defined as the temperature, in high winds and in the shade, of a bulb doused in water to allow maximum heat transfer from evaporation.

103 Weitzman, *Fat-Tailed Uncertainty*, 10.

104 Weitzman, *Fat-Tailed Uncertainty*, 10.

105 There is considerable mathematical theory behind how these functions behave and how to interpret things like the area under the curve, how to measure the width of the curve, etc. The precise details are not relevant here because the point I am making is that there are any number of ways to generate those rules. What's important is the basic idea that the PDF provides the probability (the odds) of some event (like climate sensitivity being 5°C/9°F, not 2.6°C/4.7°F). The odds are given as a percent probability and depend

crucially on the shape of the chosen curve and the distance away from the best estimate.

106   Formally, the bell curve drops away from the central tendency exponentially (faster), while the Pareto drops away polynomially (slower). Exponential rates (best exemplified by the general formula $y = e^x$ and encountered in the real world as compound interest rates) are much faster than polynomial rates (like $y = x^2$).

107   Wietzman, *GHG Targets*, 6.

108   It takes the form of *Damage (as a function of Temperature)* $= 1 / [1 + (T/x)^2]$, where x is a (somewhat) arbitrary "scaling factor" in degrees, calculated from a bunch of inputs to provide intuitively reasonable damage amounts for a very narrow range / low range of temperature increases. In DICE, x = 20.46ºC (36.83ºF).

109   To be fair, Nordhaus has already ruled catastrophic climate disruption out of court, so he isn't on the hook for making his damage function take it into account. The point I'm making is more general than this particular example.

110   Martin Parry et al., "Assessing the Costs of Adaptation to Climate Change: a Review of the UNFCCC and Other Recent Estimates," *International Institute for Environment and Development*, http://www.iied.org/pubs/pdfs/11501IIED.pdf.

111   By "global" I mean that it affects everything within some system of analysis. A "local" variable would only affect sub-sections of an analytical model.

112   Weitzman, *Fat-Tailed Uncertainty*, 11.

113   It doesn't matter whether you think this will actually happen, it's hypothetical. Pick any disaster—asteroid strike, mutant virus, whatever.

114   This is not a reasonable assumption since the fishery will undoubtedly be smaller, since we're fishing empty oceans. Still, the central lesson holds.

115   To obtain the value of a future income stream, we must first figure out the net present value (NPV) in 2100 of an income stream of $85 billion at a given discount rate. Then, we must discount that amount over a hundred years to bring it into today's money. For a one percent discount rate, the NPV in 2100

of an $85 billion income stream (over 20 years) is $1.5 trillion. Discounting to today's dollars gets $554 billion.

116   For a four percent discount rate, NPV in 2100 of the $85 billion income stream (over 20 years) is $1.15 trillion, discounted to today's dollars it's $23 billion.

117   Nordhaus, *A Question of Balance*, 5.

118   Nordhaus, *A Question of Balance*, 36.

119   Lomburg, *Cool It*, 100.

120   Lomburg, *Cool It*, 78.

121   Lomburg, *Cool It*, 70.

122   I've never really thought there was much to be envious of science since it seemed clear to me each camp did very different things. For psychology to be envious of science was like music being envious of medicine — it just made no sense. Comparing apples to oranges, as the saying goes.

123   "Nobel Winners in Economics the Reluctant Celebrities," *New York Times*, December 4, 2011.

124   See R. Kuttner, "The Poverty of Economics," *Atlantic Monthly*, February 1985.

125   To be fair to Nordhaus, he openly admits the limitations of DICE. He's an honest academic. It's those who interpret it that appear guilty of hubris.

126   Krugman, Paul, "How Did Economists Get it so Wrong?" *The New York Times*, September 2, 2009.

127   I. Basen, "Economics Has Met the Enemy and It Is Economics," *The Globe and Mail*, October 15, 2011.

Chapter 5: The Fossil Fuel Party

128   Almost all transport fuels and more than two-thirds of electrical production are derived from fossil fuels.

129   Thomas Hobbes, *Leviathan*, XIII, 9 (1651).

130   I do not claim all of us live well. Social inequity remains a profound problem.

But by and large, most of us live very well compared to life before fossil fuels.

131 North American homes use an average of one kilowatt of electricity twenty-four hours a day, all year round. A relatively fit person sweating it out on a treadmill can put out maybe a quarter of that.

132 See Tom Butler and George Wuerthner, *Energy* (Foundation for Deep Ecology in collaboration with Watershed Media and Post Carbon Institute, 2012), 2.

133 I'm just giving a sense of it. There are fine books that go into the nitty-gritty of a job like this. I recommend George Monbiot, *Heat* (Doubleday Canada, 2006); Lester Brown, *Plan B* (Norton, 2009); and the ongoing "Jacobson Plan" by Mark Jacobson & Mark Delucchi (as initially articulated in "A Path to Sustainable Energy by 2030," *Scientific American*, November, 2009).

134 An Olympic pool is around 2.5 million liters (660,000 gallons), or 26,000 barrels. North America uses 9 billion barrels of oil a year.

135 See *U.S. Billion-Ton Update: Biomass Supply for a Bioenergy and Bioproducts Industry*, United States Department of Energy, 2011. This amount may increase with harvesting of ocean-derived algae.

136 A typical capital cost of next-generation biofuels is about $2.50 per liter ($10 per gallon) per year of capacity. A $500 million plant makes 189 million liters (50 million gallons), or 1 million barrels, a year. North America uses 9 billion barrels a year; replacing a third means producing 3 billion barrels annually. We'll need three thousand of these plants.

137 One way the U.S. managed to build a very impressive war machine was to ban the sale or production of passenger vehicles.

138 Energy intensity is a measure of economic activity produced per unit of energy.

139 That the economy *must* keep growing is based on two ideas, as far as I can tell. First, capital had better get a return or it's withdrawn. The economy gets unstable when it flattens out because investors pull back their capital. Second, government debt is sustainable only if the economy grows faster than the interest rate.

140 This is optimistic, but not impossible. Converting North American gaso-
line-powered vehicles to electric drivetrains would double total electrical use,
but that's assuming the same old efficiency. So double the efficiency, keep
half the cars as gas-electric hybrids, make better use of off-peak load, and
we're almost there.

141 Total U.S. coal-based electrical capacity is 225 gigawatts (GW), or 2 billion
megawatt hours (MWh), so we need 667 million MWh for each renewable.
A 3 MW turbine at thirty-three percent capacity gets 8,760 MWh per year, so
we would need 76,000 of them. A 10 MW solar field at twenty-five percent
capacity gets 22,000 MWh per year, so we would need 30,000 of those. This
leaves us with a need for 150 giant half-GW geothermal plants.

142 The irony of storing it in old oil fields is you bury one piece of carbon to get
at another. You're not really any further ahead. The CO_2 is used to squeeze
out more oil which gets burned, which releases CO_2 . . .

143 See V. Smil, *Energy Myths and Realities* (AEI Press, 2010), 91–93.

144 Smil, *Energy Myths and Realities*, 42.

145 V. Smil, *Energy Transitions* (Praeger, 2010), 70.

146 When Al Gore talks about an energy revolution that can take place in a
decade, his scale of ambition is laudatory. But in skipping over how hard it
would be, and having us believe it's something we can make happen without
real sacrifice, it seems to me he's just making soothing noises to the already
converted.

147 A straw man refers to putting an idea or argument forward merely to deflect
attention away from the real issue. The argument is unreal, a straw man.

148 J. Rubin, *The End of Growth* (Random House Canada, 2012), 6.

149 Heffernan, *Willful Blindness*, 110.

150 R. Suskind, *Confidence Men* (Harper Collins, 2011), 100.

151 Heffernan, *Willful Blindness*, 92.

152 *United States of America v. Kenneth L. Lay*, United States District Court for
the Southern District of Texas, Houston Division, May 2006.

153 Heffernan, *Willful Blindness*, 87.

154   J. Magee and F. Milliken, *Power Differences in the Construal of Crisis: Hurricane Katrina and the 9/11 Attacks*, http://citation.allacademic.com/meta/p_mla_apa_research_citation/3/0/7/4/2/pages307425/p307425-1.php.

155   Heffernan, *Willful Blindness*, 169–170.

156   Heffernan, *Willful Blindness*, 91.

157   Galbraith, *The Affluent Society*, 7.

158   Galbraith, *The Affluent Society*, 16.

159   J.K. Galbraith, *The Culture of Contentment* (Houghton Mifflin, 1992), 6.

160   Galbraith, *The Culture of Contentment*, 81.

161   Galbraith, *The Culture of Contentment*, 6.

162   See "Greenspan — I was wrong about the economy. Sort of," *The Guardian*, October 24, 2008.

163   Galbraith, *The Culture of Contentment*, 53.

164   Galbraith, *The Affluent Society*, 11.

165   *International Herald Tribune*, October 14, 2011.

CHAPTER 6: WAKING THE FROG AND TURNING DOWN THE HEAT

166   But it's close. Our predetermined warming from past emissions is nearly 2°C (3.6°F).

167   Walmart has been effectively tackling energy efficiency, and CEO Lee Scott takes climate change seriously. As Walmart puts pressure on their suppliers, entire industries can be forced to fall into line.

168   This is given as a range — typically 2–4.5°C (3.6–8.1°F) — and is distributed over a probability curve, like the bell curves we saw in Chapter Four. As more empirical data comes in and models include more feedback effects, the range of this curve narrows. A 3°C (5.4°F) midpoint is conservative.

169   I'm assuming a symmetrical probability curve, so the midpoint of the climate sensitivity bell curve has equal areas (total probabilities) above and below that number.

170   I'm using the unit CO_2e, which is a measure of the carbon equivalent of all greenhouse gases, including methane, etc., rather than CO_2. At the time of writing, CO_2 is 396 ppm, and CO_2e is above 430 ppm.

171   Global emissions of 2.5 ppm CO_2e is equivalent to around 50 Gt (55 billion tons) in absolute terms. See N. Stern, *The Global Deal* (Public Affairs, 2009), 40.

172   Emissions rose a record 5.9 percent in 2010 after a small drop of 1.4 percent in 2009 due to the global recession. The last decade has seen an average increase of 3 percent.

173   Past absorption rates might not be maintained and may already be declining (see Galen McKinley et al., "Convergence of Atmospheric and North Atlantic CO_2 Trends on Multidecadal Timescales," *Nature Precedings*, June 8, 2011). One day our warming oceans might start *releasing* CO_2, like a can of cola in the sun. The extra CO_2 will make the ocean more acidic and hostile to much ocean life.

174   I invite the reader to play with some simple spreadsheets to drive this point home.

175   See Weitzman, *Fat-Tailed Uncertainty*, 278.

176   See Stern, *The Global Deal*, 26.

177   These proposals include how to balance emissions reductions between the galloping developing world and the developed world, the vast cost difference between early and late emissions reductions, the role efficiency can play, the relationship of economic growth and absolute energy consumption, the evolving technology landscape, etc.

178   *World Energy Outlook 2011: Executive Summary*, International Energy Agency, 2

179   J. Rojely, et al., "Probabilistic cost estimates for climate change mitigation." *Nature*, 2013, vol. 493, 80.

180   See Stern, *The Global Deal*, 26, or almost any other climate science paper on the topic.

181   J. Rogelj et al., 2013, 79.

182   According to the IEA, we must leave two-thirds in the ground, but this

makes big assumptions about changing agricultural practices and the like. Among others, 350.org put the figure at eighty percent: "It's simple math: we can burn less than 565 more gigatons of carbon dioxide and stay below 2°C [3.6°F] of warming—anything more than that risks catastrophe for life on Earth. The only problem? Fossil fuel corporations now have 2,795 gigatons in their reserves, five times the safe amount." See www.350.org.

183 J. Hansen, *Storms of My Grandchildren* (Bloomsbury USA, 2009).

184 Some qualification is required, since, as we saw in Chapter 3, there has often been a subtle and beneficial interplay between the public and private sectors.

185 By not having to pay the appropriate cost of the risk, financial firms were effectively able to ignore it.

186 The U.S. Senate recently approved $60.4 billion to help New York and New Jersey recover from Hurricane Sandy.

187 This is the "black magic" part of cap and trade. The target and rate of reduction are set by a blend of necessity (science) and possibility (industry-specific best practices). Fierce intersectoral competition for permits can be expected. If the tar sands are given too many, then the forestry sector will have to work extra hard and vice versa.

188 Renewable portfolio standards are a local regulation that forces utilities to hit some pre-defined percentage for renewable energy in their generation portfolio.

189 FITs show the ups and downs of alternative policies. A FIT is a long-term price for renewable energy. It's set to ensure a decent rate of return, and so it depends on the price of equipment at the time of contracting. FITs (mainly from Germany and Spain) launched today's global wind and solar industries. The solar industry grew enough that prices have come down three-quarters since 2008. That success came at a cost. Spain and Germany locked in expensive long-term contracts, and Ontario's FIT shows how unstable government-led processes can be. After tempting offshore wind developers with the promise of attractive rates of return, the Ontario government arbitrarily cancelled the tariff. The amount of solar allowed onto the grid keeps changing,

which makes it hard for manufacturers to plan ahead.

190 The marginal social cost is the cost to society of emitting one extra unit of emissions. A cost-benefit analysis of climate change is a bit of a black art, as we saw in Chapter 4.

191 This is the estimate of Canada's Carbon Engineering, which is developing a pilot project that uses caustic soda to extract atmospheric CO_2, which is then injected into old oil fields.

192 Rogelj et al., *Probabilistic cost estimates for climate change mitigation*, 79.

193 Stern, *The Global Deal*, III.

194 In Canada, the small but growing cleantech sector spends more in absolute terms on R&D than the much larger Canadian oil and gas sector. Globally, private R&D has dropped in half since the 1990s, to around $4 billion by the early 2000s.

195 Canada alone has earmarked almost $4 billion in provincial and federal funding for CCS projects.

196 France and Russia both have next-generation breeder reactors, but they are not being scaled to commercial relevance.

197 The IEA has pegged this number at around $60 billion a year.

198 Stern, *The Global Deal*, 112.

199 Stern, *The Global Deal*, 113.

200 Fusion's a bit of a wild card since it's been decades away for decades now. Maybe South Korea's recent billion-dollar bet with the U.S. Department of Energy will pay off, or perhaps even Canada's own General Fusion will surprise us with commercial success.

201 A 500 MW power plant operating at thirty-five percent efficiency needs 1,429 MW of energy input. One watt is equal to 1 Joule/second. If coal contains 37 megajoule/kg (about 17 MJ/pound), we need 3.337 metric tons (3.678 tons) per day, or 1.2 million metric tons (1.3 million tons) per year.

202 One km3 (1/4 cu. mile) of rock releasing just 1°C (1.8°F) of heat provides the equivalent of 70,000 metric tons (77,000 tons) of coal (see Rand, *Kick the Fossil Fuel Habit*, 70); 10°C (10°F) gives 700,000 metric tons (770,000

tons) — so a couple of cubic kilometers of rock is significant. The system is rotated through the ground beneath the coal plant as heat is extracted. Since the holes go down many kilometers, it's easy to see how a cone of modest size can capture on the order of a hundred cubic kilometers of rock.

203   My team at Action Canada produced comprehensive recommendations on this proposal back in 2009. See http://www.greenbonds.ca/GB_Policy.pdf for details and www.greenbonds.ca for a cool video.

204   Whether it is any longer or not, I don't care. I'm quite happy for someone to knock us off our pedestal.

205   A new company out of Vancouver — Fenix Energy — can retrofit geothermal systems into large condos and office towers by driving a drilling outfit into the bottom level of the parking garage. It operates in 2.2 metres (7'2") of headroom, is fully electric, and is self-contained. You no longer need space to go geothermal.

206   The discount is somewhere between twenty dollars and forty dollars per barrel.

207   Only policies already in place by mid-2012 are assumed.

208   These subsidies include $3.5 trillion for renewable energy generation plus about half a trillion for biofuels. (*World Economic Outlook 2013*, Washington, DC: International Monetary Fund, 2013, 233–236).

209   Carbon has different costs in different regions. In the Organization for Economic Co-operation and Development countries, the cost rises to $120/ton by 2035 from less than $20 now. In most other parts of the world, prices kick in around 2020 and rise more slowly to around $40. An important effect of carbon pricing is to enable CCS, which might happen around $40/ton. Let's assume a global average of $40/ton (in all areas, including those without a price) over the entire period and average global emissions of 25 Gt (27 billion tons) annually (they're just over 30 Gt/33 billion tons now and drop to 22 Gt/24 billion tons in the scenario by 2035). That costs $1 trillion per year.

210   "Achieving the Efficient World Scenario would give a boost to the global economy of $18 trillion over the Outlook period, with a 0.4% higher global

GDP in 2035 than in the New Policies Scenario. . . . This reflects a gradual reorientation of the global economy, as the production and consumption of less energy-intensive goods and services frees up resources to be allocated more efficiently. The reduction in energy use and the resulting savings in energy expenditures increase disposable income and encourage additional spending elsewhere in the economy." From *World Economic Outlook 2012* (Washington, DC: International Monetary Fund, 2012), 313.

211   The U.S. Senate never ratified the Kyoto Protocol. It was thought at the time that Al Gore's aggressive commitments were never more than feel-good posturing, given the impossibility of Senate ratification.

212   Canada was the first country that signed the Kyoto Protocol to formally opt out. While Canada claimed it was unable to meet its targets, and therefore was to face punishing financial sanctions, that was never the case. The protocol was so toothless that countries could give notification of non-compliance, and the financial penalties were non-binding.

213   Over the long term, of course, economic advantage and moral leadership merge since a low-carbon economy is the only economy that can sustain wealth. One could argue Europe has acted on that understanding, and not for moral reasons.

Bibliography

Beinhocker, E.D. *The Origin of Wealth*. Harvard Business School Press, 2006.

Blackmore, Susan. *The Meme Machine*. Oxford University Press, 1999.

Brown, L.R. *Plan B 4.0: Mobilizing to Save Civilization*. Norton, 2009.

Butler, T. and G. Wuerthner, eds. *Energy*. Post Carbon Institute, 2013.

Cartwright, N. *The Dappled World*. Cambridge University Press, 1999.

Cartwright, N. *How the Laws of Physics Lie*. Oxford University Press, 1983.

Clark, A. *Associative Engines*. MIT Press, 1993.

Clark, A. *Being There: Putting Brain, Body and World Together Again*. MIT Press, 1997.

Dreyfus, H. *What Computers Still Can't Do: A Critique of Artificial Reason*. MIT Press, 1999.

Dyer, G. *Climate Wars*. Random House Canada, 2008.

Flannery, T. *Here on Earth*. HarperCollins, 2010.

Galbraith, J.K. *The Affluent Society.* Houghton Mifflin, 1998.

Galbraith, J.K. *The Culture of Contentment.* Houghton Mifflin, 1992.

Hamilton, C. *Requiem for a Species.* Earthscan, 2010.

Hansen, J. *Storms of My Grandchildren.* Bloomsbury USA, 2009.

Heffernan, M. *Willful Blindness: Why We Ignore the Obvious at Our Peril.* Anchor Canada, 2011.

Hertsgaard, M. *Hot: Living Through the Next Fifty Years on Earth.* Houghton Mifflin Harcourt, 2011.

Hulme, M. *Why We Disagree About Climate Change.* Cambridge University Press, 2009.

Kahneman, D. *Thinking, Fast and Slow.* Farrar, Straus and Giroux, 2011.

Kauffman, S. *Reinventing the Sacred.* Basic Books, 2008.

Lakoff, G. *The Political Mind.* Penguin, 2009.

Lewis, C.S. *Selected Literary Essays.* Cambridge University Press, 1979.

Lind, M. *Land of Promise: An Economic History of the United States.* HarperCollins, 2012.

Lomborg, B. *Cool It.* Vintage, 2010.

Lovelock, J. *The Revenge of Gaia.* Penguin, 2006.

Lovelock, J. *The Vanishing Face of Gaia: A Final Warning.* Penguin, 2009.

Monbiot, G. *Heat: How to Stop the Planet from Burning.* Doubleday Canada, 2006.

Newell, P., and M. Paterson. *Climate Capitalism: Global Warming and the Transformation of the Global Economy.* Cambridge University Press, 2010.

Nordhaus, W. *The Climate Casino.* Yale University Press, 2013.

Nordhaus, W. *A Question of Balance: Weighing the Options on Global Warming Policies.* Yale University Press, 2008.

Norgaard, K.M. *Living in Denial: Climate Change, Emotions, and Everyday Life.* MIT Press, 2011.

Rand, T. *Intuition as Evidence in Philosophical Analysis: Taking Connectionism Seriously*. University of Toronto, 2008.

Rand, T. *Kick the Fossil Fuel Habit: 10 Clean Technologies to Save Our World*. EcoTen Publishing, 2009.

Rogelj, J., D. McCollum, A. Reisinger, M. Meinshausen, and K. Riahi. "Probabilistic Cost Estimates for Climate Change Mitigation." *Nature*, vol. 493 (2012): 79–83.

Schlefer, J. *The Assumptions Economists Make*. Harvard University Press, 2012.

Schneider, S., and K. Kuntz-Duriseti. "Uncertainty and Climate Change Policy." In *Climate Change Policy: A Survey*, 53–87. Island Press, 2002.

Sherwood, S.C., and M. Huber. "An Adaptability Limit to Climate Change Due to Heat Stress." *Proceedings of the National Academy of Sciences*, vol. 107, no. 2 (2010): 9552–9555.

Smil, V. *Energy: Myths and Realities*. AEI Press, 2010.

Smil, V. *Energy Transitions: History, Requirements, Prospects*. Praeger, 2010.

Speth, J.G. *The Bridge at the End of the World*. Yale University Press, 2008.

Stern, N. *The Global Deal: Climate Change and the Creation of a New Era of Progress and Prosperity*. Public Affairs, 2009.

Stiglitz, J.E. *Globalization and Its Discontents*. Norton, 2003.

Suskind, R. *Confidence Men: Wall Street, Washington, and the Education of a President*. HarperCollins, 2011.

Taleb, N. *The Black Swan: The Impact of the Highly Improbable*. Random House, 2010.

Varela, F., E. Thompson, and E. Rosch. *The Embodied Mind*. MIT Press, 1993.

Weitzman, M. "Fat-Tailed Uncertainty in the Economics of Catastrophic

Climate Change." *Review of Environmental Economics and Policy*, vol. 5, issue 2 (2011): 275–292.

Weitzman, M. "GHG Targets as Insurance Against Catastrophic Climate Damages," *Journal of Public Economic Theory*, Association for Public Economic Theory, vol. 14 (2) (2012): 221–244.

Yergin, D. *The Prize: The Epic Quest for Oil, Money & Power*. Simon and Schuster, 2009.

Yergin, D. *The Quest: Energy, Security, and the Remaking of the Modern World*. Penguin Press, 2011.

Index